Writing Scientific Research Articles

T0340740

Writing Scientific Research Articles

Strategy and Steps

Third Edition

Margaret Cargill BA, DipEd, MEd(TESOL), DEd
School of Agriculture, Food and Wine
The University of Adelaide
Adelaide
South Australia 5005
Australia

Patrick O'Connor BSc, PhD
School of Biological Sciences
School of Professions
The University of Adelaide
Adelaide
South Australia 5005
Australia

WILEY Blackwell

This edition first published 2021
© 2021 John Wiley & Sons Ltd

Edition History
© 2013 by Margaret Cargill and Patrick O'Connor
© 2009 by Margaret Cargill and Patrick O'Connor

All rights reserved. No part of this publication may be reproduced, stored in a retrieval system, or transmitted, in any form or by any means, electronic, mechanical, photocopying, recording or otherwise, except as permitted by law. Advice on how to obtain permission to reuse material from this title is available at http://www.wiley.com/go/permissions.

The right of Margaret Cargill and Patrick O'Connor to be identified as the authors of this work has been asserted in accordance with law.

Registered Offices
John Wiley & Sons, Inc., 111 River Street, Hoboken, NJ 07030, USA
John Wiley & Sons Ltd, The Atrium, Southern Gate, Chichester, West Sussex, PO19 8SQ, UK

Editorial Office
9600 Garsington Road, Oxford, OX4 2DQ, UK

For details of our global editorial offices, customer services, and more information about Wiley products visit us at www.wiley.com.

Wiley also publishes its books in a variety of electronic formats and by print-on-demand. Some content that appears in standard print versions of this book may not be available in other formats.

Limit of Liability/Disclaimer of Warranty
The contents of this work are intended to further general scientific research, understanding, and discussion only and are not intended and should not be relied upon as recommending or promoting scientific method, diagnosis, or treatment by physicians for any particular patient. In view of ongoing research, equipment modifications, changes in governmental regulations, and the constant flow of information relating to the use of medicines, equipment, and devices, the reader is urged to review and evaluate the information provided in the package insert or instructions for each medicine, equipment, or device for, among other things, any changes in the instructions or indication of usage and for added warnings and precautions. While the publisher and authors have used their best efforts in preparing this work, they make no representations or warranties with respect to the accuracy or completeness of the contents of this work and specifically disclaim all warranties, including without limitation any implied warranties of merchantability or fitness for a particular purpose. No warranty may be created or extended by sales representatives, written sales materials or promotional statements for this work. The fact that an organization, website, or product is referred to in this work as a citation and/or potential source of further information does not mean that the publisher and authors endorse the information or services the organization, website, or product may provide or recommendations it may make. This work is sold with the understanding that the publisher is not engaged in rendering professional services. The advice and strategies contained herein may not be suitable for your situation. You should consult with a specialist where appropriate. Further, readers should be aware that websites listed in this work may have changed or disappeared between when this work was written and when it is read. Neither the publisher nor authors shall be liable for any loss of profit or any other commercial damages, including but not limited to special, incidental, consequential, or other damages.

Library of Congress Cataloging-in-Publication Data is applied for

ISBN 9781119717270

Cover Design: Wiley
Cover Image: © AzmanJaka/E+/Getty Images

Set in 10.5/13pt JansonTextLTStd by SPi Global, Pondicherry, India

SKY8C1F009C-FEFD-4E4D-A630-05AED84BE3D7_051221

Contents

Preface to the third edition

The second edition of *Writing Scientific Research Articles: Strategy and Steps* has continued to be used widely, by authors and by those supporting them to develop their articles and their article writing skills. However, there have been numerous changes in the international journal publishing landscape since it was released. The online evolution of science publishing has continued, with consequences for all aspects of the submission, publication, and promotion of published work. The changes brought about by online publishing and digital sharing present some new challenges but also deepen the need for understanding of the basics of communication in this genre. This third edition responds to the major changes and incorporates further suggestions from readers and from colleagues in many places who have used the book in their teaching and mentoring of novice authors. Some new sections and exercises have been added to develop user skills – we hope they are useful!

We have included new material in Chapter 12 on review articles, focusing on systematic reviews; in Chapter 11 on visual abstracts and highlights; and in Chapter 7 on publishing Methods papers. In Chapter 5, you will find new material on visualising results, handling supplementary material, and archiving data.

To enhance its usability by its many audiences, we have listed suggested pathways through the book at the end of Chapter 1: for students in the preparation phase before writing a manuscript; for researchers with data ready to start writing their manuscript; for authors using English as an additional language (EAL); for scientists instructing or mentoring students or junior colleagues; and for language professionals teaching science research students or providing advice on draft manuscripts. An enhanced reference list provides access to recent published work that has informed the updates we've made.

Once again we express our gratitude to all who have contributed to this third edition, and especially to our editors at Wiley, Rosie Hayden and Julia Squarr. As ever, remaining errors and omissions are down to us!

Margaret Cargill
Patrick O'Connor
April 2021

Preface to the second edition

The first edition of *Writing Scientific Research Articles: Strategy and Steps* has been taken up with enthusiasm worldwide, both by novice authors themselves and by those who help prepare them for the publishing component of a science career in the 21st century. This second edition incorporates suggestions from users, additional insights we have gained in teaching from the book, and several additional sections designed to extend the book's approach to some topics not previously covered. Firstly, we have incorporated an additional article structure in Chapter 2 – one frequently used in fields such as physics, computer science, and some types of engineering – and an additional provided example article that uses it, from the field of remote sensing. These additions mean that the book now covers the full range of macro-structures commonly used in scientific research articles, extending its usefulness across a wider range of discipline areas.

The second addition is a chapter on the writing of review articles. Here, we apply the principles set out in the first edition to the challenge of writing a review article suitable for publication in an international journal. We suggest that most of the advice remains completely applicable if the term "data" is re-conceived as the author's evaluations of the work being reviewed, and the article's "take-home message" is new synthesis or conclusions that advance understanding of the field in question.

We take a similar approach for the third new feature, one that has been requested by many readers – a chapter on the writing of applications for grant funding. Although the specific requirements of funding bodies differ, the underlying process of understanding and responding effectively to a set of criteria remains the same. We have focused on applying the principles from the rest of the book to provide guidelines and strategies that will be relevant in contexts ranging from small grants for travel or conference attendance all the way to large national or international funding opportunities.

Once again we express our thanks to colleagues who have contributed to the developments included in the second edition, especially Holly Slater, Andrew Smith, John Harris, Peter Langridge, Matt Gilliham, and Michelle Picard, and to our editors at Wiley-Blackwell, Ward Cooper, Carys Williams, and Kelvin Matthews. We also thank the many users of the first edition whose ideas and questions have spurred us on. As before, any remaining problems are our own.

Margaret Cargill
Patrick O'Connor
September 2012

Preface to the first edition

Writing Scientific Research Articles is designed for early-career researchers in the sciences: those who are relatively new to the task of writing their research results as a manuscript for submission to an international refereed journal, and those who want to develop their skills for doing this more efficiently and successfully. All scientists are faced with pressure to publish their results in prestigious journals and all face challenges when trying to write and publish. This book takes a practical approach to developing scientists' skills in three key areas necessary for success:

- developing strategy: understanding what editors and referees want to publish, and why;
- developing story: understanding what makes a compelling research article in a particular discipline area; and
- using language: developing techniques to enhance clear and effective communication with readers in English.

The skills required for successful science writing are both science- and language-based, and skill integration is required for efficient outcomes. We are an author team of a scientist and a research communication teacher who have combined our perspectives and experience to produce an integrated, multidisciplinary approach to the task of article writing.

We have written the book both for those who write science in English as their first language and those for whom English is an additional language (EAL). Although a very high proportion of the research articles published worldwide currently appears in English, scientific research is an intensely international and intercultural activity in the twenty-first century, and authors come from a wide range of language and cultural backgrounds. This situation adds another layer to the challenges facing authors themselves, journal editors and referees, and those who teach and support EAL scientists. We hope the book will be relevant to all professionals involved with the practice of research article writing.

The book is designed for use either by individuals as a self-study guide, or by groups working with a teacher or facilitator. Readers can prepare their own manuscript step by step as they move through the book, or use the book as a preparation phase and return to relevant parts when the time comes to write their own paper and navigate the publishing process.

The book has arisen out of fruitful collaborations at the University of Adelaide over many years, and especially out of our work with the Chinese Academy of Sciences since 2001. There are many people to thank for their contributions both to the approach and the book. First on the language end of the continuum must be Robert Weissberg and Suzanne Buker, whose 1990 book *Writing Up Research: Experimental Research Report Writing for Students of English* laid such an effective foundation in using the insights of the worldwide community of genre-analysis researchers as the basis of effective teaching about research article writing. Next are John Swales and his colleagues over the years, for their research output, their teaching texts, and their modeling of humble and rigorous curiosity as an effective way into the worlds of other disciplines. Then the team at Adelaide that has built from these bricks a context where the book could emerge: especially Kate Cadman, Ursula McGowan, and Karen Adams, and so many scientists over the years. For bringing the perspective and experience of scientists, particular thanks go to those who have taught with us in China: Andrew Smith, Brent Kaiser, Scott Field, Bill Bellotti, Anne McNeill, and Murray Unkovich. We also thank those who have supported the training programs where we have refined our practical teaching approach, particularly Yongguan Zhu and Jinghua Cao. And, of course, the many early-career authors, in Australia, Vietnam, Spain, and China, who have participated in our workshops and contributed their insights and enthusiasm to the development of the book.

Our warm thanks go also to the people who have helped with the production of the book itself: Sally Richards, Karen Adams, Marian May, and our editors at Wiley-Blackwell, Delia Sandford and Ward Cooper. Remaining errors and omissions must be down to us.

Margaret Cargill
Patrick O'Connor
September 2008

A framework for success

CHAPTER 1

How to use this book

1.1 Getting started with writing for international publication

This book is for all authors who want improved strategies for writing effective scientific papers in an efficient way, including those new to the task. The focus is on writing in English, but many of the strategies are equally effective for writing science in other languages. Plurilingual authors – those using English as an additional language (EAL) – will find their situations and needs addressed alongside those of authors with English as a first language (EL1), as well as those common to both groups.

In this book, we will use other terms as well as *paper* for what you are aiming to write: it may be called a *manuscript*, a *journal article*, or a *research article*. (See Chapter 2 for comments on other types of scientific articles, Chapter 12 for writing review articles, and Chapter 18 for how to apply the book's approach to writing funding grant proposals.) All of these terms are in use in books and websites providing information and advice about this type of document: this *genre*. The concept of genre is important for the way this book works, as we have based our approach in writing it on the findings of researchers who work in the field of genre analysis. These researchers study documents of a particular type to identify the features that make them recognisable as what they are.

One of the key concepts in use in this field of research is the idea of the *audience* for a document as a key factor in helping an author write effectively. Whenever you write any document, it is helpful to think first about your audience: whom do you see in your mind's eye as the reader of what you are writing? The idea of audience belongs as part of a "communication matrix" made up of four elements: *audience* (as described in the previous sentence), *purpose* (what do you want the document to achieve?), *format* (how will the required format constrain how you write the document?), and *assessment* (what criteria will be used to decide if the document is successful?). We will use all the elements of this matrix to guide our discussion of the genres we will analyse in the book, and we begin now by thinking about the audience for a scientific research article.

Who is your audience?

Often the audience that you think of first is your scientific peers – people working in areas related to yours who will want to know about your results – and this is certainly a primary audience for a research article. However, there is another "audience" whose requirements must be met before your peers will even get a

Writing Scientific Research Articles: Strategy and Steps, Third Edition.
Margaret Cargill and Patrick O'Connor.
© 2021 John Wiley & Sons Ltd. Published 2021 by John Wiley & Sons Ltd.

chance to see your article in print: the journal editor and reviewers (also called referees; see Chapters 3, 13, and 14 for more information). These people are often thought of as gate-keepers (or as a filter), because their role is to ensure that only articles that meet the journal's standards and requirements are allowed to enter or pass through. Therefore, it can be useful from the beginning to find out and bear in mind as much information as you can about what these requirements are. In this book, we refer to these requirements as reviewer criteria (see Chapters 3 and 14 for details), and we use them as a framework to help unpack the expectations that both audiences have of a research article written in English. We aim to unpack these expectations in two different but closely interrelated ways – in terms of:

- the content of each article section and its presentation; and
- the English language features commonly used to present that content.

To do this, the book uses an interdisciplinary approach, combining insights from experienced science authors and reviewers about content with those from specialist teachers of research communication in English about the language. Elements of language that are broadly relevant to most readers of the book will be discussed in each chapter. In addition, Chapter 17 focuses on ways in which users of EAL can develop the discipline-specific English needed to write effectively for international publication. This chapter can be studied at any stage in the process of working through the book, after you have completed Chapter 1.

1.2 Publishing in the international literature

If you are going to become involved in publishing in the international literature, there are a number of questions it is useful to consider at the outset: Why publish? Why is it difficult to publish? What does participation in the international scientific community require? What do you need to know to select your target journal? How can you get the most out of publishing? We will consider these questions in turn.

Why publish?

We have already suggested that researchers publish to share ideas and results with colleagues. Other reasons for publishing include

- to leave a record of research which can be added to by others;
- to receive due recognition for ideas and results; and
- to attract interest from others in the area of research.

However, there are two additional reasons that are very important for internationally oriented scientists:

- to receive expert feedback on results and ideas; and
- to legitimise research; that is, to receive independent verification of methods and results.

These reasons underscore the importance of the review process we discussed earlier. However, there are difficulties associated with getting work published – difficulties that operate for all scientists, plus some that are specific to scientists working in contexts where English is a foreign or additional language.

Why is it difficult to publish?

In addition to any language-related barriers that spring to mind, it is also important to realise that writing is a skill, whatever the language. Many of the points covered in this book are equally important for EAL and EL1 scientists. In addition, because most science research contexts are now multilingual and multicultural wherever they are located, an overt focus on the role of language in writing for publication will benefit all players, from novices to mentors.

Getting published is also a skill: not all writers are published. Some reasons for this fact include the following:

- not all research is new or of sufficient scientific interest;
- experiments do not always work – positive results are easier to publish; and
- scientific journals have specific requirements which can be difficult to meet – publishing is a buyer's market.

These issues will be addressed as you proceed through the book.

Another reason that researchers find the writing and publication process difficult is that communicating your work and ideas opens you up to potential criticism. The process of advancing concepts, ideas, and knowledge is adversarial, and new results and ideas are often rigorously debated. Authors facing the blank page and a potentially critical audience can find the task of writing very daunting. This book offers frameworks for you to structure your thinking and writing for each section of a scientific article and for dealing with the publishing process. The frameworks provided will allow you to break down the large task of writing the whole manuscript into small tasks of writing sections and subsections, and to navigate the publishing process.

What does participation in the international scientific community require?

A helpful image is to think about submitting a manuscript to an international journal as a way of participating in the international scientific community. You are, in effect, joining an international conversation. To join this conversation, you need to know what has already been said by the other people conversing. In other words, you need to understand the "cutting edge" of your scientific discipline: what work is being done now by the important players in the field internationally. This means:

- getting access to the journals where people in the field are publishing;
- subscribing to the e-mail alert schemes offered by journal publishers on their websites so that you receive tables of contents when new issues are published; and
- developing effective skills for searching the Internet and electronic databases to which you have access.

Without this understanding, it will be difficult to write about your work so as to show how it fits into the progress being made in your field. In fact, this knowledge is important when the research is being planned, well before the time when the paper is being written: you should try to plan your research so it fits into a developing conversation in your field.

Active involvement in international conferences is an important way to gain access to this international world of research in your field. Therefore, you need confident skills in both written and spoken English for communication with your peers. This book aims to help with the written language as used in international journals, and some ideas for developing spoken science English are given in Chapter 16. As you become a member of the international research community in your field in these ways, you will develop the knowledge base you need to help you select the most appropriate journal for submission of your manuscript: we call this your *target journal*.

What do you need to know to select your target journal?

Choosing the right journal for your manuscript will influence the chance of getting published relatively easily and quickly. You should be thinking about the journal you want to publish in from the beginning of your research, and should have made a choice by the time you begin to write the Introduction and Discussion sections of your paper.

The right journal for you is the journal which optimises the speed and ease of publication, the professional prestige you accrue, and the access for your desired audience. These factors are interwoven, and it can be helpful to develop a publication plan to maximise your publication success. The journal of your choice may not choose to accept your article, and you are advised to have a list of preferred journals to turn to if you are rejected from your first choice. Here, we set out some issues to consider when choosing a journal for your manuscript:

- Does the journal normally publish the kind of work you have done? Check several issues and search the journal website. It is helpful if you can cite work from the journal in the Introduction of your manuscript, to show that you are joining a conversation already in progress in the journal. Examine some of the key articles you refer to in your Introduction, and check which journals are cited in the Introductions of these articles. By following back through the literature, you should be able to develop a mind-map of the journals in the field of your research. The journals that are most often cited in the Introduction and Discussion sections of your manuscript will be most likely to accept work in your field.
- Do the aims and scope of the journal match the content and the level of impact of your work for the field? Check the websites or issues of potential journals to identify those with scope and aims most appropriate for your manuscript. In this way, you can try to ensure that your article will reach the audience you want to read it, once it is published.
- Is the journal of an appropriate standard for your needs? First, does it referee its papers? This is absolutely imperative for enhancing the international credibility of your work. It may also be important to check the journal's impact factor, if this measure is important for assessing research outcomes in your country or research context. (See Appendix for more information on impact factor, citation index, and other similar measurements.)
- Does the journal publish reasonably quickly? Many journals include the dates when a manuscript was received and published underneath the title information, so you can check the likely timeline. Others include this information on their websites. Journals which publish online versions of papers before the print version will usually have a faster time to publication. Journals want to publish submissions

quickly to ensure they attract authors who are doing innovative and new work. You may also want to publish your research quickly to ensure that others do not publish similar work before you, and to increase your publication and citation record for promotions and grants.

- Are there charges associated with publishing in the journal? Some journals charge authors a fee to publish, or to publish coloured illustrations. Check whether this is the case. If so, you can ask whether the journal is willing to waive these charges for authors in some parts of the world. You may also want your research to be accessible to a wide range of readers who do not have access to libraries or other subscriptions to journals in your field. Many journals now offer to provide Open Access to papers (i.e. to make them accessible for free download without subscription to the journal) if the authors pay an upfront fee. Check whether the journal of your choice offers this service if you want (or are required by your institution or funding body) to pay for Open Access.
- Are members of the journal's editorial staff efficient and helpful? For example, some journals have information on their website with targeted advice for authors from EAL backgrounds. You may be able to ask colleagues who have submitted to particular journals about their experiences. It can be useful to share this kind of information among colleagues in your laboratory group or work team, perhaps as part of a programme to encourage international publication of the work of your institution or group.

How can you get the most out of publishing?

Publishing quickly is often helpful. In addition, publishing in a widely read journal is better for you (higher citation index; see Appendix). However, if you aim too high in relation to the international value of the work you have done, you may be rejected, and resubmission takes more time. These two issues have to be balanced carefully to determine an optimal strategy for your particular situation. Finally, publishing where your peers will read your paper is important. Use Task 1.1 to summarise your understanding and complete Table 1.1. Discuss the outcomes with your co-authors to develop your publication strategy.

Task 1.1   Analysing potential target journals

To optimise the outcomes from publishing your manuscript, we recommend that you develop a publishing strategy. Part of this strategy is to select the best journal for submission. In order to make this choice, first select three or four preferred journals in your field that you think would accept your manuscript. Then answer the following questions for each one and record the answers in Table 1.1:

1 Has the journal published similar work with a similar level of novelty to yours in the last 3 years? Record a "yes" or "no" (if "no," think carefully before submitting your manuscript to this journal).

2 Do the journal's scope and the contents of recent articles match the main components of your manuscript, in terms of subject, methods, and results? (Write down the main type of papers published, e.g. "plant physiology: non-molecular studies").

3 What is the measure of relative journal quality/impact which is most important to you and your field of research? Record the score or measure for each journal (e.g. journal impact factor or journal cited half-life).

4 What is the journal's time to publication? (This may be on the journal's website or recorded for articles in the journal.) Record the time or a score for fast or slow (e.g. less than 3 months from acceptance = fast; more than 1 year = slow).

5 Does the journal have page charges or provide Open Access if you want it? (And can you pay if payment is required?)

Examine the journal scores you have recorded in Table 1.1 and rank your preferred journals in order of overall preference, taking all criteria into consideration.

Table 1.1 Rating preferred journals in terms of key criteria for maximising your publication success.

Journal name	Recent publication of similar work and novelty	Match of scope and recent content to your work	Journal quality/ impact	Time to publication	Page charges or Open Access costs
1					
2					
3					
4					

This publication strategy should include making appropriate use of relevant forums or networks, both local and international and both in-person and via social media, to publicise your article and its findings upon publication – but the first step remains getting published in your selected target journal.

Once you have thought about the issues raised here, and made some preliminary decisions about a possible target journal, you are ready to move on to consider the aims of this book.

1.3 Aims of the book

The aims of the book are to provide you, the reader, with:

- an improved understanding of the structure and underlying logic of scientific research articles published in English in the international literature;
- an overall strategy for turning a set of results into a paper for publication;
- skills for analysing the structure and language features of scientific articles in your own discipline, and for using the results of this analysis to improve your own scientific writing;
- knowledge of the stages involved in the process of submitting an article for publication, and strategies for completing each stage;
- knowledge and basic mastery of the specific English language features commonly used in each section of published articles;

- strategies and tools for improving your drafts, such as structured checklists, ways to strategically re-use relevant language elements, special-purpose software, and discipline-specific writing groups; and
- a process for completing a draft of an article on your research results, prepared in the style of the journal to which you wish to submit.

1.4 How the book is structured

Two principles underlie the way we have organised this book: that people learn best by doing; and that you will want to continue developing your skills on your own or with colleagues in the future, even if you first encounter the book in a classroom environment. Therefore, we aim to show you how you can use examples of journal articles, from your own field and from others, to learn more about writing for publication.

To achieve this goal, the book will often invite you to discuss examples with a colleague and then report to a larger group. This assumes that you are using the book in a class situation. However, if you are using it for individual study, you can note down your answers and then revise them once you reach the end of the section. As we move through the book, you will also have the opportunity to draft (or substantially revise) your own article, section by section, if this is appropriate.

Instructions for activities in the book will use the following terms to refer to different categories of example articles:

- **Provided Example Articles (PEAs)**: these are three articles chosen by the authors of the book and included in full at the back (Chapters 19, 20, and 21). You will use all three in the early sections of the book and then be asked to select one to use in more detail.
- **Selected Article (SA)**: this is an article that you will choose from your own field of research and that may be from your target journal. Use Task 1.2 to choose your SA.
- **Own Article (OA)**: this is the draft manuscript you will write using your own results as you progress through the book. If you do not yet have your own results, you can skip the tasks relating to the OA and come back to them later.

Task 1.2   Selecting an article to analyse

Select an article in your own field of research to use as your Selected Article (SA), preferably from your target journal and preferably written by a first-language speaker of English (check authors' names and the location of their work sites to help identify their language background). We suggest that you do not choose your SA from *Nature* (UK) or *Science* (USA), as they use conventions that are very different from those of most other journals – it will be more useful to learn the standard conventions first, and then adapt them later if you need to. (See Chapter 2 for more details on the differences in article structure.)

The following sections of the book work like this.

- We present information about the structure of research articles, section by section, which has been summarised from the work of scholars in the field of applied linguistics over the last 20 years. We present this as a *de*scription, not a *pre*scription: that is, "this is what the scholars have found," not, "this is what you should do." We do this because there are many effective ways to write articles, not just one way. Our aim is to help you develop a repertoire (a range of effective possibilities) to select from, depending on the goals you have for a given article section.
- We then ask you to look at the relevant section of the PEAs and check whether you can find the described features there (answers to the Tasks can be found in the Answer pages at the end of the book).
- Next, we ask you to analyse your own SA for the same features and think about possible reasons for what you find.
- Finally, we ask you to work on the draft of your OA, using the new information you have gained from the analysis. (These sections are optional for readers who do not have their own results ready to write up.)
- As well as this analysis of structural features, the book includes teaching, analysis, and exercises on elements of English language usage that are particularly relevant to each section of a research article (again, solutions can be found in the Answer pages). If English is your first language, you may choose to skip some or all of these sections. However, they may be helpful when you are mentoring or co-authoring with colleagues from EAL backgrounds – a common requirement of 21st-century science workplaces.
- After all the sections of a research article have been covered in this way, we demonstrate how to apply the same principles to writing a review article for submission to an international journal.
- We then focus on the process of submitting the manuscript to the journal, and how to engage in correspondence with the editor about possible revisions.
- Chapter 15 summarises a process for preparing a manuscript from first to last, using the knowledge built up through your engagement with previous chapters on each article section, and including strategies for editing and checking.
- Chapter 16 focuses on techniques and strategies for ongoing development of your skills for writing, publishing, and presenting your research in English.
- Chapter 17 provides advice about specific features of science writing that often cause problems for EAL authors. It can be studied at any stage of a reader's progress through the book.
- Chapter 18 considers the writing of proposals for research funding and shows how to apply the book's strategic approach to this important component of 21st-century science.
- Section 5 contains the three PEAs (Chapters 19, 20, and 21).
- At the end of the book, you will find Answers to the tasks that appear in the other chapters, an Appendix containing details of journal impact assessment, and the Reference list.

1.5 How to use this book if you are. . .

. . .a research student getting ready to write a manuscript

- After reading Chapter 1 you will have a good idea of where to go next. The logic of the book follows the chapter numbers.

- Chapter 3 is essential reading for what follows.
- All of Section 2 comes next, but Chapter 12 may be left till it is relevant for you.
- Chapter 15 sets you up to begin writing.
- Chapter 16 contains ideas for ongoing skill development; implement the ones most relevant to your situation, including setting up a writing group if appropriate.
- Chapters 13 and 14 give valuable background to the task ahead in terms of submitting and the review process.

…a research student or early-career researcher drafting your manuscript

- Revisit the parts that contain advice specific to your stage of writing, making sure that Chapters 2 and 3 are understood as a basis.
- Chapters 4–11 give you the base you need to use the process set out in Chapter 15 as a commencing strategy (plus Chapter 12 if you are writing a review paper).
- Consult Chapter 13 in detail before the drafting is completed, and Chapter 14 both before submission and again once the reviews arrive.

…an author using English as an additional language (EAL)

- Over and above the advice relevant to your stage of writing, Chapter 1 has introduced you to our approach to developing discipline-specific English skills as an integral part of writing for publication training.
- Chapter 17 can be read at any stage of your engagement with the book. We especially recommend Section 17.4; our experience over many years indicates that EAL authors report positive outcomes when they have taken the time to construct their own discipline-specific research article corpus and learned to use concordancing software.

…a scientist instructing or mentoring research students or junior colleagues

- The book can be provided to students and early-career researchers as a stand-alone resource, and strategies discussed in Chapter 16 can be used to enhance its effectiveness.
- Promising outcomes have been reported when a scientist uses the book as the basis of a Writing Group programme for research students, in an Australian context of no coursework in research degree programmes (Cargill & Smernik 2016).
- In a Chinese context, collaborative input from a scientist into a course taught from an English for Research Publication Purposes perspective has been analysed, and ways in which it can be more than a "cameo" role identified (Li et al. 2019). Collaborative, interdisciplinary teaching of the book's contents is highly recommended (Cargill & O'Connor 2010).

…a language professional teaching science research students or providing advice on draft manuscripts

- Throughout, the book provides language professionals with a thorough introduction to the specifics of science research article writing from a content perspective, while integrating language concerns at every stage.

- The approach described in Section 17.4 is a useful tool when providing advice on draft manuscripts, using concordancing software with a discipline-specific corpus of research articles. It is also useful for identifying discipline-specific examples for teaching purposes.
- Chapter 17 provides pointers to some English language features that have been identified as regularly troublesome to EAL authors.
- The book, or selected parts of it, can be used as the basis of courses for research students where the degree structure includes coursework (Cargill et al. 2018), or of workshops where it includes only research (Cargill 2019). Collaborative interdisciplinary teaching of the material by language specialists and scientists has been shown to be highly effective (Cargill 2011).

CHAPTER 2

Research article structures

We will now look at the overall structure of research articles in science. In general, this follows a set of conventions that have developed over the years from 1665, when the first issue of *Philosophical Transactions* appeared in England. It is important to recognise that, within a common core structure, there are variations from field to field and from journal to journal: always check the specific requirements of your target journal before finalising the structure of any article you write.

Before we look at the results of research into article structure, complete Task 2.1.

Task 2.1   Article headings and subheadings

Read quickly to find the headings of the sections of the Provided Example Articles (PEAs) (Chapters 19, 20, and 21):

- How is each paper organised?
- What are the main headings and subheadings? Make brief notes.

Check your answers in the Answer pages.

  Now look at the headings of your Selected Article (SA) and the SA of a colleague. Note the similarities and differences you find.

2.1 Conventional article structures: AIMRaD and its variations

Before we explore article structure in detail, it is important to note that our main focus in this book is on research articles based on experimental research (but see Chapter 12 for a discussion of review articles). Other research paradigms, for example in humanities and social science fields, use somewhat different structures for their papers. But even within the sciences, there is variation in the structures preferred or allowed by particular journals for presenting different types of research studies in various fields of research. First, we will consider the hourglass diagram (Figure 2.1) commonly used to represent the structure of an AIMRaD (Abstract, Introduction, Materials and methods, Results, and Discussion) article, arguably the most commonly used structure, and what it can tell us about English language research articles. In this diagram, it is the width

Writing Scientific Research Articles: Strategy and Steps, Third Edition.
Margaret Cargill and Patrick O'Connor.
© 2021 John Wiley & Sons Ltd. Published 2021 by John Wiley & Sons Ltd.

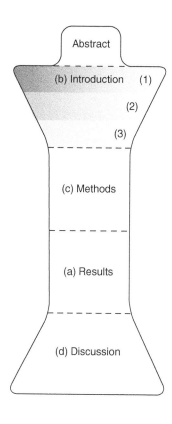

(a) The whole structure is governed by the Results box; everything in the article must relate to and be connected with the data and analysis presented in the Results section.

(b) (1) The Introduction begins with a broad focus. The starting point you select for your Introduction should be one that attracts the lively interest of the audience you are aiming to address: the international readers of your target journal.

(3) The Introduction ends with a focus exactly parallel to that of the Results; often this is a statement of the aim or purpose of the work presented in the paper, or its principal findings or activity.

(2) Between these two points, background information and previous work are woven together to logically connect the relevant problem with the approach taken in the work to be presented to address the problem.

(c) The Methods section, or its equivalent, establishes credibility for the Results by showing how they were obtained.

(d) The Discussion begins with the same breadth of focus as the Results – but it ends at the same breadth as the starting point of the Introduction. By the end, the paper is addressing the broader issues that you raised at the start, to show how your work is important in the "bigger picture."

Fig. 2.1 AIMRaD (Abstract, Introduction, Materials and methods, Results, and Discussion): the hourglass "shape" of a generic scientific research article and key features highlighted by it.

and shape of the segments, rather than their depth, that tell us something important about scientific articles.

Here we represent an experimental article in terms of different component shapes put together into an hourglass configuration. This enables us to highlight several important features of such articles in a way that is easy to remember. The right-hand part of Figure 2.1 summarises the features to focus on at this stage.

Of course, not all scientific research articles follow the simple structure given in Figure 2.1. There are three major variations that we will introduce here; these are presented visually in Figures 2.2, 2.3, and 2.4. Study these figures now, before doing Task 2.2.

Task 2.2   Structure of the PEAs

Check the notes you made in answer to Task 2.1.

- Which of the four structures presented in Figures 2.1–2.4 matches most closely the structures of the PEAs? Check your answer in the Answer pages.
- Which most closely matches your SA?

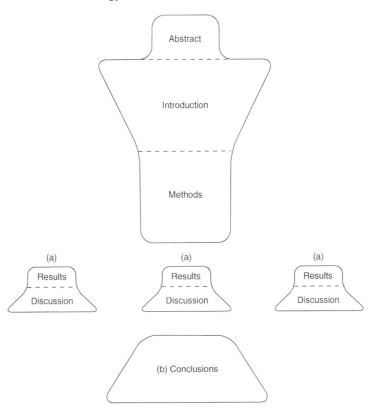

(a) The Methods section, often renamed Procedure or Experimental, is presented after the Discussion, sometimes in a smaller typeface than the rest of the paper.

(b) This change means that more details may need to be given in the Results section to explain how the results were obtained.

Fig. 2.2 AIRDaM (Abstract, Introduction, Results, Discussion, and Methods and materials): a structure variation that occurs in articles in some journals, notably in the fields of chemistry and molecular biology.

(a) The Results and Discussion are presented together in a single combined section; each result is presented, followed immediately by the relevant discussion.

(b) This change means that a separate section is needed at the end to bring the different pieces of discussion to ether; it is often headed Conclusions.

Fig. 2.3 AIM(RaD)C (Abstract, Introduction, Materials and methods, repeated Results and Discussion, Conclusions): a structure variation that is permitted in some journals, usually for shorter articles.[1]

[1] Variations of this arrangement are also found. For example, *Advanced Electronic Materials* uses the following headings for full papers: Abstract (unlabeled), Introduction, Results and Discussion, Conclusion, Experimental Section. This represents a combination of features seen in the AIRDaM and AIM(RaD)C structures presented in this chapter. It is therefore very important to check the structure of recent articles in the journal to which you wish to submit.

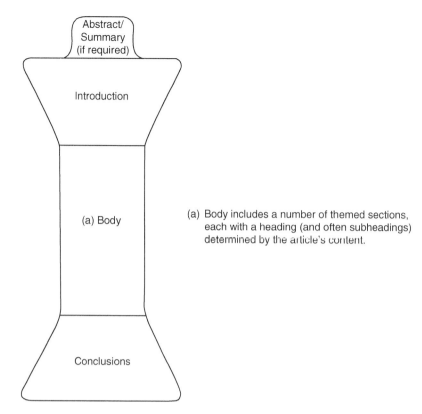

Fig. 2.4 AIBC (Abstract, Introduction, Body sections, Conclusions): a structure variant common in fields such as some kinds of engineering, computer science, remote sensing, and physics. Note that body sections may contain various combinations of theory, methods, results, and discussion and that the headings and subheadings used for the sections depend on the research "story" being told rather than a set formula. *Source*: Burgess, S., & Cargill, M. (2013). Using genre analysis and corpus linguistics to teach research article writing. *Supporting Research Writing*, 55–71. doi:10.1016/b978-1-84334-666-1.50004-7. © 2013, Woodhead.

Other research article formats

The highly cited journals *Nature* (UK) and *Science* (USA) use variations of the common conventions for their article categories, reflecting the fact that their aim is to present highly significant new advances in science in ways that are very accessible to scientists who are not necessarily specialists in the areas covered by the articles. These articles typically begin with a carefully structured initial section introducing the background and rationale of the work to the wide range of expected readers, followed by a concise report of the findings and a short discussion. Methods are often only summarised in the main article, with full details appearing on a linked website. Details of the structures required by these journals can be found on their websites. Competition for publication in these journals is intense, and they are not likely to be realistic targets for most beginning scientists. For this reason, we do not focus on their structure in this book.

Many journals offer alternatives to the full article format for reporting research findings. Important among these are brief notes (also called *research notes* or *notes*) and letters. These may not include any section headings at all, but if you read them with an analytical eye you will be able to find the same types of information as are contained in a full article, arranged according to whichever structure is most appropriate for the study and the journal conventions.

Now we begin to think in more detail about what information appears in the different sections of a research article. It is likely that you already know quite a lot about this, from reading articles for your own work. Task 2.3 focuses on this pre-existing knowledge.

Task 2.3   Prediction

Identify which part of a research paper the following phrases came from. Write one of the following letters at the end of each line: (**I**) = Introduction, (**M**) = Materials and methods, (**R**) = Results, or (**D**) = Discussion (remembering that in the AIBC structure some of these functions may appear under different headings).

It is very likely that . . . because . . .	(**D**)
. . . yielded a total of . . .	( )
The aim of the work described . . .	( )
. . . was used to calculate . . .	( )
There have been few long-term studies of . . .	( )
The vertical distribution of . . . was determined by . . .	( )
This may be explained by . . .	( )
Analysis was carried out using . . .	( )
. . . was highly correlated with . . .	( )

Check your answers in the Answer pages.

It is likely that the clues you use to help you answer the questions in Task 2.3 relate both to the vocabulary in the phrases and to elements of the grammar, especially the tense of the verbs (simple past, present perfect). We will build on this knowledge in later sections.

In Chapter 3, we will consider the relationship between the structure of research articles and the expectations of the gatekeeper audience that you, as an article submitter, are aiming to meet. The conventional structures we have been looking at in this chapter have been maintained in science journals for a long time; we can assume that they serve the purposes of the journal editors effectively and meet the needs of the journal readers. It is interesting to think about how and why that is the case.

CHAPTER 3

Reviewers' criteria for evaluating manuscripts

As discussed in Chapter 1, the first audience for your manuscript is the editor of the journal you have selected. In recent years, with the advent of electronic submission by uploading files on a computer, the very first audience may be a person who checks that formatting and other requirements have been met, but this fact does not alter the editor's initial filtering role in terms of the article's content. If the manuscript is judged suitable for reviewing (see Chapters 13 and 14 for more details on this process), the editor sends it to (usually) two peer reviewers or referees for comment. These reviewers are probably working in the same field as you; perhaps their names are in your list of references. However, the reviewing process is "blind," meaning that you do not know who reviews your paper. (Double-blind reviewing, where the referees also do not know who authored the manuscript they are reviewing, is less commonly practised in the sciences.)

Each journal has its own set of instructions for reviewers, and sometimes these are available on the journal's website. You should check and see whether this is the case for the journal you are targeting, and obtain a copy if possible. For the purposes of this book, we have constructed a composite list of reviewer criteria that includes the sorts of questions referees are commonly asked to respond to (Figure 3.1). In addition to "ticking the boxes" by providing yes/no answers to these questions, reviewers are asked to write comments about any problems with the manuscript or any suggestions they have for improvements that need to be made before the manuscript can be considered suitable for publication in the journal. Increasingly, as the number of manuscripts submitted to journals grows, reviewers are asked to give some numerical rating of the paper's novelty or quality as well (e.g. "Does this manuscript fall within the top 20% of manuscripts you have read in the last 12 months?"). Reviewers return their comments to the editor. To help you think more about these criteria, complete Task 3.1 now.

As we discuss each section of a research article in detail, we will keep these reviewer criteria in mind and draw attention to the presentation features and English expressions that are commonly used to highlight the fact that evidence relevant to reviewer criteria is being presented.

We will begin by considering the question: Does the title clearly indicate the content of the paper?

Writing Scientific Research Articles: Strategy and Steps, Third Edition.
Margaret Cargill and Patrick O'Connor.
© 2021 John Wiley & Sons Ltd. Published 2021 by John Wiley & Sons Ltd.

1. Does the contribution fall within the scope of the journal?
2. Is the contribution new and original?
3. Is the contribution of sufficiently high impact for the journal?
4. Does the manuscript demonstrate an adequate awareness of relevant research?
5. Are the interpretations and conclusions sound and justified by the data presented?
6. Is the organisation of the manuscript acceptable?
7. Does the title clearly reflect the contents?
8. Is the abstract sufficiently informative, especially when read in isolation?
9. Does the introduction set the manuscript in an international context and show how it builds on previous work?
10. Do the methods and analysis conform to acceptable standards and is sufficient detail presented to comprehend the study?
11. Are all the figures and tables necessary and the captions adequate and informative?
12. Is the length of the manuscript appropriate to the content?
13. Are the references adequate for the subject and length of the manuscript?
14. Is the quality of the English satisfactory?

Fig. 3.1 Typical questions that referees are asked to answer when reviewing manuscripts for science journals.

Task 3.1   Where would reviewers look?

Read the list of questions in Figure 3.1. For each, decide where in a manuscript a reviewer would expect to find evidence on which to base their answer. Write one or more of the following abbreviations beside each question: T (title), A, I, M, R, D, or Ref (reference list) (see Chapter 2). For example, for question 5, you would write *M and R*. Again, bear in mind that in an AIBC (Abstract, Introduction, Body sections, Conclusions) structure the elements M, R, and D may appear in sections with other names. Check your answers in the Answer pages.

3.1 Titles as content signposts

Good titles clearly identify the field of the research, indicate the "story" the results tell, and raise questions about the research in the mind of the reader. We will return to a more detailed consideration of titles in Chapter 10. For now, consider this example and then complete Task 3.2.

Title: Bird use of rice field strips of varying width in the Kanto Plain of central Japan

Information:
The focus is on birds in relation to rice fields.
The width of rice field strips was varied in the study.
Width of strips was correlated with the number and species of birds using them.
The research took place in central Japan.

Possible questions:
Why was the width of the strips an important variable?
Did the width of the rice field strips affect which birds used it?
If so, which field strip width was used most by which birds?

How did the researchers measure bird use?
Would the experiment be worth repeating for rice field strips in other places?

21

Reviewers' criteria for evaluating manuscripts

Ch 3
Reviewers'
criteria for
evaluating
manuscripts

Choosing one of the example articles as your focus for analysis tasks

Titles B, C, and D in Task 3.2 are the titles of the Provided Example Articles (PEAs) included at the back of the book. You will need to select one of them to use as the basis of text analysis exercises as we proceed through the sections of the book. The answers you gave to the questions in Task 3.2 should help you to decide which of the three articles will be more interesting and relevant to you.

Finally, complete Task 3.3.

Task 3.2   Information extracted from titles

Look at the following titles and list the information about the research and its results you can deduce from each. What questions might you, as a reader, expect to answer by reading the article? (The questions will depend on the individual reader's reason for reading the text.)

Title A: Use of *in situ* ^{15}N-labelling to estimate the total below-ground nitrogen of pasture legumes in intact soil-plant systems

Information:

Questions:

Title B: Short- and long-term effects of disturbance and propagule pressure on a biological invasion

Information:

Questions:

Title C: The soybean NRAMP homologue, GmDMT1, is a symbiotic divalent metal transporter capable of ferrous iron transport

Information:

Questions:

Title D: An emergent strategy for volcano hazard assessment: from thermal satellite monitoring to lava flow modeling

Information:

Questions:

Check your answers with the suggestions provided in the Answer pages.

Task 3.3   Unpacking the title of your Selected Article (SA)

Repeat Task 3.2 for the title of your SA.

Title:

Information:

Questions:

When and how to write each article section

CHAPTER 4

Results as a "story": the key driver of an article

Because the results govern the content and structure of the whole article, it is important to be as clear as possible about the main points of your results "story" at the beginning of the writing process. We suggest that your first task when preparing to write a paper is to identify from your results a clearly connected story which leads to one or more "take-home messages." This term refers to what readers remember after they have put the paper down: what they talk to their colleagues about over a cup of coffee next day, for example.

To move towards this clear story, we recommend "mapping out" the manuscript by producing a set of dot points on the main results, with comments on them and their meanings. The manuscript mapping process is described in more detail in Chapter 15, but it draws on what you will learn about the components of each section of a manuscript in the chapters that follow. The key to mapping out your story is to focus on your tables and figures (and text results) and, for each one, to write a list of one or two bullet points highlighting the main message(s) of the data presented. Next to each point, indicate whether it contributes to the Results or Discussion section, or both. Some points may also stimulate thoughts for inclusion in the Introduction or Methods section, or be useful for the Conclusion; mark the points to indicate this using the steps described in Section 15.1. Sort the figures and tables into the best order to connect the pieces of the story together, and from the dot points you have drafted for the Results and Discussion sections, draft a take-home message and a working title (or a couple of versions of a working title). Your manuscript will be developed around the take-home message, and you should sit down with all your co-authors and discuss the story of the paper that you will write. From the starting point of the results bundles (figures, tables, and text) and dot points, aim to reach agreement with them on

- which data should be included;
- what are the important points that form the story of the paper; and
- what is/are the take-home message or messages.

This is also a good time to decide on who will be listed as authors of the paper, how the work will be allocated, and what the timelines are for different tasks (see Chapter 15 for more details). Providing some structure to the shared work plan for you and your co-authors can help everyone to prioritise the necessary work and avoid delays. This meeting is also a good time to discuss potential target journals,

Writing Scientific Research Articles: Strategy and Steps, Third Edition.
Margaret Cargill and Patrick O'Connor.
© 2021 John Wiley & Sons Ltd. Published 2021 by John Wiley & Sons Ltd.

taking into consideration the points raised in Section 1.2. Once these foundation steps are completed, you are ready to begin writing the various sections of the manuscript itself.

We have found Task 4.1 useful in helping authors identify some key information that will help them begin the drafting process.

Once you can answer these questions for your own results, you are ready to refine your tables and figures so that they present, as clearly and forcefully as possible, the data that support the components of your story. That refinement process is the topic of Chapter 5.

Task 4.1 Questions to focus the drafting process

Answer these four questions, in English even if it is not your first language, for the results you want to turn into a paper:

1 What do my results *say*? (two sentences maximum, a very brief summary of the main points, no background!)
2 What do these results mean in their context? (i.e. what conclusions can be drawn from these results?)
3 Who needs to know about these results? (i.e. who specifically forms the audience for this paper you are going to write?)
4 Why do they need to know? (i.e. what contribution will the results make to ongoing work in the field? Or, what will other researchers be missing if they have not read your paper?)

CHAPTER 5

Results: turning data into knowledge

The data presentation in a scientific article aims to illustrate the story, present evidence to support or reject a hypothesis, and record important data and meta-data. We verify, analyse, and display data to share, build, and legitimise new knowledge. To maximise the effectiveness of data visualisation, we must present all necessary data in ways which make the most important points most prominent. For a deeper understanding of the visualisation of data, the reader is referred to the work of Edward Tufte (2001, 2020), who has produced some of the most beautiful books on visualisation for research.

Data presentation is also an exercise in deciding which datasets or details to leave out of an article. If you have decided to include figures or tables, they should be numbered and presented sequentially and referred to in that order in the text. Many journals now accept additional data which support or extend the story as appendices or supplementary online data. For each data element in your paper you should ask yourself if it is necessary to the story of the paper, or not essential but valuable for those who might access it in an online archive. Remember, the referees will be asked to comment on whether all the tables and figures are necessary, and this will include the supplementary material.

Data presentation styles vary with discipline and personal preference and change over time (e.g. the visualisation of big data is becoming more common and presents a new set of challenges), and the published literature provides a large amount of contradictory advice about what to do and what looks good. Our aim in this section is not to provide a concrete set of rules for data presentation but rather to help you optimise the presentation of your data to support the story of your article. One over-arching guideline is that tables and figures should "stand alone": that is, the reader should not need to consult the text of the article to understand the data presented in them; all necessary information should appear in the table/figure itself, in the title/legend, or in keys or footnotes.

The first reference for style of data presentation is the *Instructions to Contributors* (sometimes called Instructions to Authors or Author Guidelines, or a similar name) of the journal you intend to submit the article to. Not all Instructions to Contributors provide great detail about data presentation, but they will generally guide you in formatting and preferred style. The next best source of information on data presentation style is articles in recent issues of the journal, or else of journals in the same field of similar or higher quality

Writing Scientific Research Articles: Strategy and Steps, Third Edition.
Margaret Cargill and Patrick O'Connor.
© 2021 John Wiley & Sons Ltd. Published 2021 by John Wiley & Sons Ltd.

(remembering not to stray from the Instructions to Contributors of the chosen journal). You can maximise your chances of meeting the journal's requirements by analysing the types of data presented, the choice of figures and tables, the choice of figure types, and the amount and style of data presentation in the text and in the titles and legends of articles it has published. Use the results of your analyses to inform your decisions on the data presentation for your own manuscript.

The choice of whether to use a figure, table, or text for a given dataset or results set depends on the point or meaning you want the reader to receive. Each form of data display has strengths and weaknesses, as summarised in Table 5.1.

Table 5.1 The choice between tables and figures for data display.

Most useful	Table	Figure
When working with	Numbers	Shapes or pictures
When concentrating on	Individual data values	Overall patterns
When accurate or precise actual values are	More important	Less important
When calculations and components of data are	Important	Less important
When comparisons are	Multiple, complex, important	Simple, less important

5.1 Designing figures

Design each figure to display most simply the information you want to get across most strongly. In an era when authors have access to many computer graphics packages and the ability to produce numerous graphical representations and styles, it is important to take charge of the software and direct it to your purpose. It may be helpful to determine the design elements you want in the figure before going to the graphics package. This will help you avoid using default settings or template styles which do not meet your needs. Figures can also be designed to be appealing to the eye by ensuring they have high contrast (high background to ink ratio), have proportions close to 3 : 2, and are boxed when simple and unboxed when containing numerous lines, bars, columns, or linkages.

Commonly used figure types allow you to emphasise different qualities of the data:

- **Pie or hierarchy charts** are effective at highlighting proportions of a total or whole.
- **Column and bar charts** are effective at comparing the values of different categories when they are independent of one another (e.g. apples and oranges).
- **Line charts** allow the display of a sequence of variables in time or space, or of other dependent relationships (e.g. change over time).
- **Radar charts** are useful when categories are not directly comparable.

- **Heat maps** show the magnitude of a phenomenon in a two-dimensional matrix using differences in colour (hue) or intensity.
- **Venn diagrams** allow the illustration of discrete and intersecting elements in a group of sets.

A review of figures in published articles shows a number of common strengths in best practice examples which enhance the power of figures to contribute to the communication of the story.

Best practice

- Figure types chosen are appropriate to the data and the information being conveyed, and trivial or non-existent relationships are not prominent.
- The order of data categories is sorted to highlight key information and important relationships between elements.
- Display elements (e.g. symbol, line, bar) and their qualities (e.g. style, colour, weighting) are chosen to highlight the most important information.
- Contrasting display elements (e.g. line weighting or colour or style differences) are selected to emphasise differences or similarities between elements.
- Scale, scale intervals, maximum and minimum values, and statistical representations are chosen to highlight the most important information.
- Non-sequential data are not joined in time series where there are missing values.
- Numbers are only included when the exact values are important to the story and the approximate values cannot be estimated from the axes.
- Data in the figures are not unnecessarily duplicated in the text or tables.
- Plot density is clear for large datasets (e.g. use of colour intensity or unfilled symbols).
- Data are adjusted for visualisation of very large datasets (e.g. using aggregation/simplification (e.g. box and whisker plots), jittering, binning or other methods).
- Figure styles and formats are consistent throughout the article (i.e. consistent symbols, colours and intensities, and axis ranges).
- Figures are free from clutter, and unnecessary lines or symbols are removed.
- Figures are formatted according to the Instructions to Authors for the chosen journal and consistent with the style of recent articles in that journal.
- Figures are designed to convey the necessary information at the size at which they will be displayed on the published page or screen.
- Text used in figures (e.g. legends, titles, axes) is descriptive and avoids the jargon of the research discipline or study.
- Statement titles are used to focus the viewer's attention and highlight the most important relationships or results (many of the points discussed in Chapter 10 on article titles apply to titles for figures as well).

There are other forms of figures which are not presentations of the results but which demonstrate process (e.g. flow chart), methodology (e.g. apparatus), or documentary evidence. These may have been collected originally as visual images (e.g. photographs or spatial representations). All such forms should follow the same basic rules just discussed.

To improve your own practice, it helps to examine the figures in published articles so you can draw on best practice and innovation and observe where improvements could be made. Complete Tasks 5.1 and 5.2 now.

Task 5.1 Improving information visualisation in a figure

Examine Figure 5.1 and suggest improvements which could be made based on the best practice recommendations in Section 5.1.

Check your suggested improvements against those demonstrated in the Answer pages.

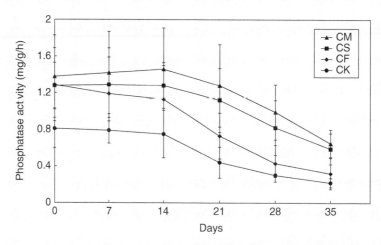

Fig. 5.1 Comparisons of root surface phosphatase activity of wheat plants for control (CK), exclusively chemical fertilizer (CF), combined application of chemical fertilizer and wheat straw (CS), and farmyard manure (CM) treatments. Error bars represent the standard error of the mean for each treatment.

Task 5.2 Examining data display

Examine your Selected Article (SA) for types of data and how they are displayed.

1 Is the overall picture or trend obvious in the way the data are presented? Could it have been made more prominent?
2 What comparisons between elements interest you, and do the presentation type and style make these comparisons easy?
3 Are the necessary details of datasets presented to allow you to make calculations from the data?
4 Do the figures omit any of the best practice features described in Section 5.1, and how do these omissions detract from the telling of the story?

5.2 Designing tables

Tables are often used to record the data and meta-data of a study. They may contain a number of rows or columns, which require careful reading before the meaning can be appreciated. This is especially true where tables contain a large number of cells and where comparisons between different rows and columns are necessary to understand the story. These potential limitations can be largely overcome by good design, particularly in terms of table layout, choice of data for inclusion, ordering of data within the table, and clear and informative row/column headings and table title. Many of the visual design elements are common to those discussed for figures: keep tables free of clutter, and define abbreviations in the title or by using footnotes. In addition, do not box tables, and use horizontal lines as separators and space to separate columns.

A review of tables in published articles shows a number of common strengths in best practice examples which enhance the power of tables to contribute to the communication of the story.

Best practice

- Data are grouped and ordered to highlight key information and important relationships between data categories.
- Contrast is used to ensure key elements stand out. Good use is made of "white space."
- Element styles are chosen to convey information and facilitate comparisons (e.g. font, bold, italics, letter case, symbols, super- or subscript).
- Unnecessary or redundant data are removed (e.g. data that are not referred to in the text and do not contribute to the story).
- Non-significant or over-precise numbers are excluded (to avoid a false sense of accuracy, or clutter).
- Data required to allow the reader to make important calculations from experimental results are included (in either the tables or the text).
- Statement titles are used to focus the reader's attention and highlight the most important relationships or results (many of the points discussed in Chapter 10 on article titles apply to titles for tables as well).

To improve your own practice, it helps to examine the tables in published articles so you can draw on best practice and observe where improvements could be made. Complete Tasks 5.3 and 5.4 now.

5.3 Figure legends and table titles

Figure legends and table titles (both also designated as *captions*) should explain what the data being presented are and highlight the key points of the part of the results story presented there. The key points of the story presented should stand alone; that is, the reader should not need to read the rest of the text to understand them. Tables and figures which effectively and clearly communicate a part of the story make the work of reviewers easy and improve the readability of articles for all users.

Figure legends have a general form with five parts. These parts usually occur in sequence, but explanation of symbols and notation (Part 5) may be interspersed in the other parts.

1 A title which summarises what the figure is about.
2 Details of results or models shown in the figure or supplementary to the figure.

Task 5.3   Improving information visualisation in a table

Examine Table 5.2 (showing data from a study using different methods of analysing potassium (K) concentration in soils with different mineralogies) and suggest improvements which could be made based on the best practice recommendations in Section 5.2.

Check your suggested improvements against those demonstrated in the Answer pages.

Table 5.2 Soil test K and mineralogy of soils (SD = standard deviation).

| Soil | Clay (g/kg) | Silt (g/kg) | mg K/kg soil | | |
			WS	CaCl$_2$	NaTPB
1	380	200	10	41	480
2	535	265	31	162	1208
3	410	230	15	57	583
4	434	205	19	70	652
5	485	235	27	100	932
6	610	282	50	290	1730
7	360	190	6	34	360
8	440	235	20	87	723
Mean	456.8	230.3	22.3	105.1	833.5
SD (±)	83.4	31.9	13.9	84.9	448.9

Task 5.4   Evaluating table design

Examine the tables in your SA or another article from a journal in your field.

1 Are all the data presented in the tables necessary, and are they sorted to make the main results most prominent?
2 Are the titles descriptive or story-telling? Could a story-telling title be written for each table?
3 Are all values calculated to the correct number of significant figures and rounded to show appropriate precision?
4 Do the tables omit any of the best practice features described in Section 5.2, and how do these omissions detract from the telling of the story?

---

3 Additional explanation of the components of the figure, methods used, or essential details of the figure's contribution to the results story.
4 Description of the units or statistical notation included.
5 Explanation of any other symbols or notation used.

Table titles can also include all of these elements but tend to have only brief Parts 2 and 3 and not to have a Part 5. Now complete Task 5.5.

Task 5.5 Identifying parts of figure legends

Read the following figure legends from the Results sections of the PEAs by Kaiser et al. (2003), Britton-Simmons & Abbott (2008), and Ganci et al. (2012), and identify the parts described in Section 5.3.

- Uptake of Fe(II) by GmDmt1 in yeast.
  (a) Influx of $^{55}Fe^{2+}$ into yeast cells transformed with GmDmt1;1, *fet3fet4* cells were transformed with GmDmt1;1-pFL61 or pFL61 and then incubated with 1 μM $^{55}FeCl_3$ (pH 5.5) for 5- and 10-min periods. Data presented are means ± SE of ^{55}Fe uptake between 5 and 10 min from three separate experiments (each performed in triplicate).
  (b) Concentration dependence of ^{55}Fe influx into *fet3fet4* cells transformed with GmDmt1;1-pFL61 or pFL61. Data presented are means ± SE of ^{55}Fe uptake to over 5 min ($n = 3$). The curve was obtained by direct fit to the Michaelis-Menten equation. Estimated K_M and V_{MAX} for GmDmt1;1 were 6.4 ± 1.1 μM Fe(III) and 0.72 ± 0.08 nM Fe(III)/min//mg protein, respectively.
  (c) Effect of other divalent cations on uptake of $^{55}Fe^{2+}$ into *fet3fet4* cells transformed with pFL61-GmDMT1;1. Data presented are means ± SE of ^{55}Fe (10 μM) uptake over 10 min in the presence and absence of 100 μM unlabelled Fe^{2+}, Cu^{2+}, Zn^{2+} and Mn^{2+}. (from Kaiser et al. 2003, Figure 5)
- Number of *Sargassum muticum* (a) recruits and (b) adults in field experiment plots (900 cm^2). Propagule pressure is grams of reproductive tissue suspended over experimental plots at beginning of experiment. The average mass of an adult *S. muticum* (174 g) is indicated by an arrow. Data are means ± 1 SE ($n = 3$). (from Britton-Simmons & Abbott 2008, Figure 1)
- A sequence of SEVIRI images recorded from 9:15 to 10:30 GMT on 13 May. At 9:15 GMT, no evident eruptive phenomena are observable. At 9:30 GMT the image shows the beginning of an ash plume (yellow pixels) moving toward NE, which was associated to the lava fountain from the northern part of the eruptive fissure. The first hot spot, due to the increased lava output and thermal anomaly, was detected by HOTSAT at 10:30 GMT. (from Ganci et al. 2012, Figure 4)

Check your answers in the Answer pages.

5.4 Supplementary material

Supplementary material is supporting material (including detailed methods, extended datasets, video files, sound clips, figures, and tables) that is not essential for inclusion or cannot be accommodated in the published manuscript but which would nevertheless aid the reader's understanding. The material should not be essential to following the conclusions of the paper, but should be relevant to the content. Each element should be referred to at an appropriate point in the manuscript in the same way that figures and tables are referred to. Any journal which allows supplementary material will specify the requirements in the Instructions to Authors or on the publisher's website. Journals may set page number or file size limits on supplementary material.

Supplementary material may not be edited by the journal and may appear online in association with the published article in exactly the format in which it was submitted. It is therefore the author's responsibility to ensure the quality is high, and figures and tables should be prepared to the standards described in this chapter and the Instructions to Authors for the journal. The supplementary material should be submitted with the article. The format for supplementary material should be accessible in any Internet browser.

5.5 Archiving data

Most quality journals require authors to demonstrate that they are archiving data for preservation and future use, because data in published papers are important products of the scientific and communication enterprise. It is usual that the journal will specify a requirement for data archiving in an appropriate public archive which offers open access and guaranteed preservation. Archiving is the author's responsibility, but the journal may set standards for archiving as part of agreeing to publish the associated manuscript.

The data must be archived in a format that allows a third party to interpret them. The British Ecological Society, for example, requires that "the archived data must allow each result in the published paper to be recreated and the analyses reported in the paper to be replicated in full to support the conclusions made. Authors are welcome to archive more than this, but not less."

Authors usually have the option to withhold availability of the archived data from the public by embargoing access for up to a year after publication.

The requirements for archiving data, and options for acceptable data archives, will usually be described in the Instructions to Authors.

CHAPTER 6

Writing about results

The Results section is the primary location of information about what is new and significant about the contribution of the paper to the field. Editors and referees will examine the results carefully to judge the scale and scope of the contribution and to check that it fits with the aims of the journal. The Results section is where visual, numerical, and textual display attracts the reader's attention to the merit of the work.

In writing sentences about their results, effective authors highlight the main points only. Published advice from editors and researchers indicates that it is important that authors *do not* repeat in words all the results from the tables or figures. This advice often suggests that authors should only write sentences about the most important findings, especially the ones that will form part of the focus of the Discussion.

6.1 Structure of Results sections

Results are sometimes presented separately from the Discussion, sometimes combined in a single Results and discussion section, and sometimes alone or combined with methods and discussion in body sections within an AIBC (Abstract, Introduction, Body sections, Conclusions) structure. Check in the Instructions to Contributors for the journal you are targeting to see which format it prefers, or examine a selection of articles if the Instructions to Contributors are not sufficiently explicit.

If the separate style is used, it is generally important to confine any comments in the Results section to saying what the numbers show, without comparing them with other research or suggesting explanations. However, authors sometimes include comparisons with previous work in the Results section where the point being made relates to a component of the results that will not be discussed in detail in the Discussion. For an example, see the first Provided Example Article (PEA), Kaiser et al. (2003), p. 152, column 2, line 7 and following.

In general, keeping Results and Discussion sections separate is more common, and we discuss only the writing of results sentences here.

Writing Scientific Research Articles: Strategy and Steps, Third Edition.
Margaret Cargill and Patrick O'Connor.
© 2021 John Wiley & Sons Ltd. Published 2021 by John Wiley & Sons Ltd.

6.2 Functions of Results sentences

The text of a Results section (and sentences written about results in the combined formats) can be generated from dot points developed around the results groups (Chapter 4) and typically

- highlights the important findings;
- locates the figure(s) or table(s) where the results can be found; and
- may comment on the results.

Elements that highlight and locate are sometimes combined in the same sentence and sometimes appear in separate sentences.

Examples of combined highlight + location styles:

Measurements of root length density (Figure 3) revealed that the majority of roots of both cultivars were found in the upper substrate layers.

The response of lucerne root growth to manganese rate and depth treatments was similar to that of shoots (Figure 2).

Example of a separate location statement:

Figure 17 shows the average number of visits per bird.

Note the different verb tenses used in the two styles. Now complete Task 6.1.

Task 6.1   Separate location sentences in Results sections

First skim (read quickly) the Results section of your selected PEA. Count how many instances of separate location sentences you find. Why do you think the authors chose to write their Results section as they did? Check your answers in the Answer pages.

Now do the same exercise for your Selected Article (SA). Discuss your findings with a colleague, if appropriate.

6.3 Verb tense in Results sections

As a first step in considering this issue, complete Task 6.2.

Task 6.2   Verb usage in Results sections

Read the following extract from a Results section and identify which *verb tenses/verb forms* are represented by the underlined words in each sentence (present, past, or modal verb). Can you think of a reason for the use of different tenses in different sentences? (N.B. The past participles used as adjectives in the passage have not been underlined, only the finite verbs.)

Antibodies <u>were raised</u> in rabbits against the N-terminal 73 amino acids of GmDmt1;1 (Figure 1c). This antiserum <u>was used</u> in Western blot analysis of 4-week-old total soluble nodule proteins, nodule microsomes, PBS proteins and PBM, isolated from purified symbiosomes. The anti GmDMT1 antiserum identified a 67-kDa protein on the PBM-enriched nodule protein fraction (Figure 3a), but did not cross-react with soluble nodule proteins, PBS proteins or nodule microsomes (Figure 3a). Replicate Western blots incubated with preimmune serum (Figure 3b) did not cross-react with the soybean nodule tissue examined. The protein identified on the PBM-enriched protein fraction is approximately 10 kDa larger than that predicted by the amino acid sequence of GmDmt1. The increase in size may be related to extensive post-translational modification (e.g. glycosylation) of GmDmt1, as it occurs in other systems. (Kaiser et al. 2003)

Summarise your findings using the following sentence starters:

- In Results sections, the past tense is used to talk about . . .
- The present tense is used in sentences that . . .
- Modal verbs are used to . . .

Compare your answers with the points listed under *Common uses of tense in Results sections*.

Common uses of tense in Results sections

- Past tense (either active or passive voice) is used when the sentence focuses on the completed study: what was done and found.
- Present tense is used

  - to describe an "always true" situation; and
  - when the sentence focuses on the document, which will always be there. N.B. Although there are no examples of this usage in the paragraph used in Task 6.2 from Kaiser et al. (2003), here is an example from McNeill et al. (1997):

    The effect of urea concentration on the fed leaf and shoot growth in subterranean clover is summarised in Table 1.

- Modal verbs (e.g. *may* and *could*) may be used in comments, especially in *that* clauses. (See Chapter 9 for more details about modal verb use in research writing.)

Now complete Task 6.3 to consolidate your learning.

Ch 6
Writing about results

Task 6.3   Analysing your SA Results section for verb usage

Choose one subsection of the Results section in your SA. Answer the following questions and discuss your findings with a colleague:

- For each verb in the subsection, why do you think the authors chose to use the tense they did?
- Do the authors use tenses in the ways discussed in *Common uses of tense in Results sections*? If not, what reasons can you suggest for this?

If you find many instances where the tense usage differs from the guidelines given in *Common uses of tense in Results sections*, we suggest that you look at two or three other papers from your field and check the tense usage in their Results sections as well. If you discover patterns that differ from our guidelines, congratulations! Make a note of your findings to guide your own future use.

Hint: Example papers from your own discipline provide the most accurate guidelines for you.

It is probably not possible to write a book that presents accurately the writing conventions of every different subfield of science. Rather than aiming to provide all the answers, we have set out to give you tools and questions to use in analysing example articles from your own research area. We want you always to check what we suggest against these examples and in this way to refine the guidelines we give so that they are as accurate as possible for the articles you need to write in order to submit to journals relevant to your field. We believe this comparison process is a valuable component of the descriptive and discovery-based method for learning about research article writing that we present in this book. Now complete Task 6.4, if appropriate.

Task 6.4   Drafting your own Results section

Begin to draft a Results section for your Own Article (OA), writing about the tables or figures you have already worked on. A good place to start is with dot points about results you have generated from your results sets and discussed with your co-authors (as described in Section 15.1).

CHAPTER 7

The Methods section

7.1 Purpose of the Methods section

Traditionally, students are taught that the Methods section provides the information needed for another competent scientist to repeat the work. In your experience of reading papers, is this what you find? Many participants in workshops we have conducted report that they have had problems in replicating what authors have done in their published studies even after reading the Methods section thoroughly.

Another way to think about the goal of the Methods section is that it establishes credibility for the results and should therefore provide enough information about how the work was done for readers to evaluate them; that is, to decide for themselves whether the results actually mean what the authors claim they mean. Referees are likely to look in this section for evidence to answer the question: Do the methods and the treatment of results conform to acceptable scientific standards?

A short note on the naming of this section of a research article is in order here. As you have seen from your analysis of the Provided Example Articles (PEAs) in Chapter 2, practice varies. Alternative names for a single section describing methods include Methods, Materials and methods, and Experimental procedures. In papers that follow the AIBC (Abstract, Introduction, Body sections, Conclusions) structure, methods may be described in more than one body section, none of which may actually be labelled Methods. For the sake of simplicity, however, we use the term Methods throughout this chapter.

It is generally accepted that methods that have been published previously can be cited and need not be described in detail, unless changes have been made to the published procedures. However, if the previous publication is not readily available to your international audience (e.g. the original journal is written in a language other than English), it is recommended that you give the details in your paper, as well as the citation to the original source. Include the language of its publication in brackets in the reference list, if appropriate. Any novel method should be described in full. You should aim to include the information a referee might need to access in order to be confident of the appropriateness and effectiveness of the method used.

Writing Scientific Research Articles: Strategy and Steps, Third Edition.
Margaret Cargill and Patrick O'Connor.
© 2021 John Wiley & Sons Ltd. Published 2021 by John Wiley & Sons Ltd.

If a goal of the Methods section is to help readers evaluate the findings presented in the Results section, then the author needs to make it clear how the two sections relate to each other. In addition, readers may come to the Methods from anywhere in the article. Burrough-Boenisch (1999) reported that 21–27% of the science editors and reviewers in her study read the Methods earlier than its place in the article structure, and 29–33% later. Two strategies can help with showing the connections:

- **Strategy 1**: Use identical or similar subheadings in the Methods and the Results sections.
- **Strategy 2**: Use introductory phrases or sentences in the Methods that relate to the aims: "To generate an antibody to GmDmt1;1, a 236-bp DNA fragment coding for 70 N-terminal amino acids was amplified using the PCR, …"

An additional strategy to clarify the logic of the Methods section is to use the first sentence of a new paragraph to introduce what you will be talking about and relate it to what has gone before. In the following example, *disturbance treatment* refers to a concept that has been mentioned previously, and the sentence introduces the reader effectively to the content of the paragraph to follow (Britton-Simmons & Abbott 2008, p. 70, paragraph 2):

> The disturbance treatment had two levels: control and disturbed. Control plots were …

In many research fields, there is a standard set of headings and a standard order for presenting Methods, so check papers in your target journal that report a similar methodology to yours. In all cases, the more detail that is included in the subheadings, the better, as this makes it easier for readers to find exactly what they need to answer their specific query about how the study was conducted. Now complete Tasks 7.1 and 7.2.

Task 7.1   Materials and methods organisation

Look at the Methods section of your selected PEA (for Ganci et al. 2012, read Section 2) and answer the following questions:

1 What subheadings are used in the section?
2 How do the subheadings relate to:

  i the end of the Introduction?
  ii the subheadings in the Results section?

3 Is the section easy for you to follow? Why? Or why not?

Compare your answers with our suggestions in the Answer pages.

  Now, repeat the task for your Selected Article (SA) and discuss your findings with a colleague or teacher, if appropriate.

Task 7.2   Planning your Methods section

For your Own Article (OA), which elements do you plan to include in the Methods section, and in what order?

7.3 Methods in supplementary material

As was discussed in Chapter 5, supplementary material is supporting matter that is not essential for inclusion or cannot be accommodated in the published manuscript, but which would nevertheless aid the reader's understanding. If including methods in the supplementary material, they should not be essential to understanding the conclusions of the paper, but should be relevant additional or complementary information. Methods in the supplementary material should be referred to in the main manuscript at an appropriate point. The target journal may specify requirements for including methods in the Instructions to Authors. Journals may set page number or file size limits on supplementary material.

7.4 Publishing methods papers

Scientific research makes advances through the development and application of research methods. Research articles in most disciplines include a Methods section to outline materials and methods used and any adaptations to established practice. There are also journals which specialise in publishing new methods as a way of promoting and facilitating the uptake of innovations within different research communities (e.g. *Methods in Ecology and Evolution*, *Chemistry Methods*, and *Biology Methods & Protocols*), or else important modifications and customisations of existing methods (e.g. *MethodsX*). Methods articles follow similar conventions to those outlined throughout this book; however, the focus is on the method itself and Introduction, Results, and Discussion sections may be absent or truncated. The journal *MethodsX*, for example, provides templates for methods and protocols which require an abstract (and graphical abstract), the method, and references (www.journals.elsevier.com/methodsx). A key element of communicating a new or modified method is presentation of evidence of the change relative to established methods and the improvement in efficiency of the new or changed method.

7.5 Use of passive and active verbs

Researchers commonly write about materials and methods in the *passive voice*: that is, using passive-voice verbs. These verb forms emphasise the action and remove emphasis from the doer of the action, but they often use more words than the corresponding active-voice verbs. Many books written to advise researchers about improving their writing recommend that they avoid the passive and use active verbs as much as possible, because this makes the writing more direct and less wordy. We agree that the passive is often over-used in science writing in general. However, we suggest that the choice is not always a simple one, especially in Methods sections. We will do the following things:

- refresh your memory on the difference between active and passive verb forms;
- consider reasons why an author might wish to choose a passive verb; and
- present some guidelines for avoiding common problems with passive verb use.

Active and passive verb forms

When we use an active verb, the grammatical subject (the answer to who or what is in front of the verb) actually does the action indicated by the verb. For example:

subject	+	active verb	+	object
The dog		**bit**		**the man.**

With a passive verb, the grammatical subject does not do the action of the verb (the biting, in this case). For example:

subject	+	passive verb	+	agent
The man		**was bitten**		**by the dog.**

The agent is often omitted in passive sentences, which is why this form is popular when the action is more important than the actor, as in many experimental procedures. Figure 7.1 summarises the difference between the two sentence constructions.

If authors of research articles are comfortable with using active-voice sentences with "we" as the subject, as in the example in Figure 7.1, then it is relatively easy to avoid the passive voice, even in Methods sections. However, many authors are not comfortable with this usage, or do not like the repetitive sound of many "we" sentences together, and passive verbs can still frequently be found in science writing.

Formation of passive-voice verbs requires an auxiliary – a part of the verb *to be* (*was* is used in the example about the man and the dog) – plus the past participle of

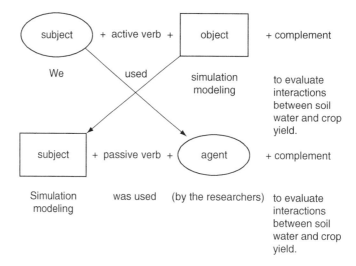

Fig.7.1 Changing an active-voice sentence to a passive-voice sentence.

Task 7.3   Active/passive sentences

Find one passive sentence from the Methods section of your selected PEA, and rewrite it in the active voice. Then find a sentence in the active voice that uses a transitive verb, and rewrite it in the passive voice. We provide some sample answers from each article in the Answer pages.

---

a verb (*bitten* in the example). Remember, only a transitive verb – a verb that has an object (indicated in dictionaries as *vt.*) – can have a passive form. Now complete Task 7.3 to consolidate your learning.

Factors influencing the choice of an active or passive verb

First, does the reader need to know who or what carried out the action? If this information is unimportant, you may choose to use a passive verb. Consider the following example:

> The researchers collected data from all sites weekly.

It is not important *who* collected the data, so the sentence may be better in the passive:

> Data[1] were collected weekly from all sites.

Second, does it sound repetitive (or immodest) to use a personal pronoun subject? For example:

> We calculated least significant differences (l.s.d.) to compare means.

This may sound more appropriate in the passive:

> Least significant differences (l.s.d.) were calculated to compare means.

Note the following points in relation to active/passive choice:

- The need to avoid repetition can explain the almost complete absence of active-voice sentences in the Experimental procedures section of the PEA by Kaiser et al. (2003): in the active, the subject of nearly every sentence would be "we."
- If you are working in a discipline where single-authored papers are common, you will need to check in a range of example papers whether it is appropriate to use "I"; in our experience, this usage is quite rare in science writing, especially in Methods sections.
- Does it help the information flow to choose either the active or passive voice?

[1]N.B. *Data* is a plural word of Latin origin, and it is still common for editors to require its use with plural verb forms. However, this convention is in the process of changing and you are likely to see it used both ways: *the data show* and *the data shows*.

In English sentences, effective writers generally connect their sentences to one another by putting old information, which the reader already knows something about, before new information (see Section 8.8 for a fuller explanation). Sometimes writers may choose a passive verb so that they can use this linking strategy. In the following example, the old information is in italics, and the active and passive verbs are identified in square brackets.

> We used [active] the results of these analyses to inform the construction of mechanistic candidate functions for the relationship between propagule input, space availability and recruitment. *These candidate functions* were compared [passive] using differences in the Akaike information criteria (AIC differences; Burnham and Anderson 2002). We then used model averaging [active]… (Britton-Simmons & Abbott 2008, p. 70)

Common problems with writing passive sentences

There is one common problem with writing passive sentences that makes them unwieldy and difficult for readers to follow. In order to make your writing easier to understand, take particular care *not* to write sentences with very long subjects and a short passive verb right at the end. For example:

> ✗ Wheat and barley, collected from the Virginia field site, as well as sorghum and millet, collected at Loxton, were used.

Instead, try to get both the subject and the verb within the first nine words of the sentence, and make sure any list of items is at the end, as in the following example:

> ✓ Four cereals were used: wheat and barley, collected from the Virginia field site; and sorghum and millet, collected at Loxton.

N.B. This improved example demonstrates a very effective sentence structure for writing lists in English. A short introduction clause (which could be a sentence on its own) is followed by a colon (:) to introduce the list. Because the two items in the list have internal commas, the items themselves are separated with a semicolon (;). This use of punctuation makes it very clear which parts of the sentence belong together and which are separated. Now complete Task 7.4.

Task 7.4 Top-heavy passive sentences

1   Here is another example of a top-heavy sentence, with a very long subject followed by a short passive verb near the end. Rewrite the sentence to make it easier for a reader to understand.

Actual evapotranspiration (T) for each crop, defined as the amount of precipitation for the period between sowing and harvesting the particular crop plus or minus the change in soil water storage in the 2 m soil profile, was computed by the soil water balance equation (Xin, 1986; Zhu and Niu, 1987). From Li et al. (2000).

Check your answer in the Answer pages.

2 Select one subsection of the Methods in your SA and check whether the authors have avoided this problem. Can you find any sentences that are difficult to follow? How could you improve them? Discuss your findings with a colleague.

Abbreviating passive sentences to avoid sounding repetitive

You may find it useful to abbreviate passive sentences, as shown in Table 7.1. Now complete Task 7.5, if appropriate.

Table 7.1 Abbreviating passive sentences to avoid excessive repetition.

Original sentence	Possible abbreviation
The data were collected and they were analysed using …	The data were collected and analysed using …
The data were collected and correlations were calculated …	The data were collected and correlations calculated …
The data which were collected at Site 1 were analysed using …	The data collected at Site 1 were analysed using …[a]

[a] In the absence of a following adverbial phrase (i.e. "at Site 1"), it would be more common to write, "The collected data were analysed."

Task 7.5   Revising your own Methods section

Use what you have learned to improve your draft of the Methods section of your OA.

CHAPTER 8

The Introduction

As your primary reading audience of editor and referees will probably start reading at the Introduction, an effective Introduction is particularly important. Referees are likely to look here for evidence to help answer the following questions:

1 Is the contribution new and original?
2 Does the contribution fall within the scope of the journal? (Meaning, this particular journal.)
3 Does the manuscript demonstrate an adequate awareness of relevant research?
4 Does the Introduction set the manuscript in an international context and show how it builds on previous work?

8.1 Argument stages towards a compelling Introduction

Applied linguistics researchers have identified a number of main stages that commonly appear in research article Introductions (Figure 8.1). These stages have been identified through analysing many published articles, and interesting variations have been found across different subdisciplines of science. However, for our purposes in this book, the broad stages presented here give us a useful framework that is flexible enough to be applicable in most contexts. But please remember that they do not represent a recipe to be followed unreflectively; rather, they provide a pattern for you to test on papers in your own field, and to refine into a useful tool for your own use.

These stages do not always occur strictly in the order given in Figure 8.1, and some may be repeated within a given Introduction. For example, Stage 2/Stage 3 sequences often recur when an author wants to justify specific aspects or components of a study. To help you see what we mean by these stages, we first ask you to read the article Introduction presented in Table 8.1 and consider our identification there of the stages and their locations. Then do Task 8.1.

Writing Scientific Research Articles: Strategy and Steps, Third Edition.
Margaret Cargill and Patrick O'Connor.
© 2021 John Wiley & Sons Ltd. Published 2021 by John Wiley & Sons Ltd.

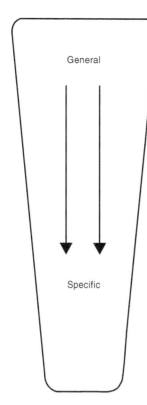

General

Specific

(1) Statements about the field of research to provide the reader with a setting or context for the problem to be investigated and to claim its centrality or importance.

(2) More specific statements about the aspects of the problem already studied by other researchers, laying a foundation of information already known.

(3) Statements that indicate the need for more investigation, creating a gap, a need for extension, or a research niche for the present study to fill.

(4) Very specific statements giving the purpose/objectives of the writer's study or outlining its main activity or findings.

(5) Statement(s) that give a positive value or benefit for carrying out the study (optional).

(6) A "map": statements telling how the rest of the article is presented (only in some research fields).

Fig. 8.1 Argument stages of an Introduction to a science research article. *Source*: Modified from Weissberg, R. & Buker, S., *Writing Up Research: Experimental Research Report Writing for Students of English*. Prentice Hall Regents, Englewood Cliffs. © 1990 Pearson.

Table 8.1 Identification of stages in the Introduction to "Use of *in situ* ^{15}N-labelling to estimate the total below-ground nitrogen of pasture legumes in intact soil-plant systems" (McNeill et al. 1997, reproduced by permission of CSIRO Publishing).

Extract	Stage
Current estimates of the below-ground production of N by pasture legumes are scarce and rely mainly on data from harvested macro-roots (Burton 1976; Reeves 1984) with little account taken of fine root material or soluble root N leached by root washing. Sampling to obtain the entire root biomass is extremely difficult (Sauerbeck and Johnen 1977) since many roots, particularly those of pasture species (Ellis and Barnes 1973), are fragile and too fine to be recovered by wet sieving. Furthermore, the interface between the root and the soil is not easy to determine and legume derived N will exist not only as live intact root but in a variety of other forms, often termed rhizodeposits (Whipps 1990).	Stage 1 Stage 3 in "scarce" and "little account" Stage 1
An approach is accordingly required which enables *in situ* labelling of N in the legume root system under undisturbed conditions coupled with subsequent recovery and measurement of that legume N in all of the inter-related below-ground fractions.	Stage 3 (broad gap)
Sophisticated techniques exist to label roots with ^{15}N via exposure of shoots to an atmosphere containing labelled NH_3 (Porter et al. 1972; Janzen and Bruinsma 1989) but such techniques would not be suitable for labelling a pasture legume within a mixed sward.	Stage 2 Stage 3
Labelled N_2 atmospheres (Warembourg et al. 1982; McNeill et al. 1994) have been used to label specifically the legume component of a mixed sward via N_2 fixation in nodules.	Stage 2

(*Continued*)

Table 8.1 (*Continued*) 49

The Introduction

Extract	Stage
However, these techniques require complex and expensive enclosure equipment, which limits replication and cannot be easily applied to field situations; furthermore, non-symbiotic N_2 fixation of label may occur in some soils and complicate the interpretation of fate of below-ground legume N.	Stage 3
The split-root technique has also been used to introduce ^{15}N directly into plants by exposing one isolated portion of the root system to ^{15}N either in solution or soil (Sawatsky and Soper 1991; Jensen 1996), but this necessitates some degree of disturbance of the natural system. Foliar feeding does not disturb the system and has the additional advantage that shoots tolerate higher concentrations of N than roots (Wittwer et al. 1963). Spray application of ^{15}N-labelled urea has been successfully used to label legumes *in situ* under field conditions (Zebarth et al. 1991) but runoff of ^{15}N-labelled solutions from foliage to the soil will complicate interpretation of root-soil dynamics. Russell and Fillery (1996), using a stem-feeding technique, have shown that *in situ* ^{15}N-labelling of lupin plants growing in soil cores enabled total below-ground N to be estimated under relatively undisturbed conditions, but they indicated that the technique was not adaptable to all plants, particularly pasture species. Feeding of individual leaves with a solution containing ^{15}N is a technique that has been widely used for physiological studies in wheat (Palta et al. 1991) and legumes (Oghoghorie and Pate 1972; Pate 1973). The potential of the technique for investigating soil-plant N dynamics was noted as long as 10 years ago by Ledgard et al. (1985) following the use of ^{15}N leaf-feeding in a study of N transfer from legume to associated grass. The experiments reported here were designed (1) to assess the use of a simple ^{15}N leaf-feeding technique specifically to label *in situ* the roots of subterranean clover and serradella growing in soil, and (2) to obtain quantitative estimates of total below-ground N accretion by these pasture legumes.	Stage 2 Stage 3 Stage 2 Stage 3 Stage 2 (Stage 3 implicit in "potential") Stage 3 Stage 2 Stage 4 (aims of the present study)

Source: McNeill, A.M., Zhu, C.Y., & Fillery, I.R.P., Use of in situ 15N-labelling to estimate the total below-ground nitrogen of pasture legumes in intact soil-plant systems. *Australian Journal of Agricultural Research*, 48, 295–304. © 1997 CSIRO Publishing.

Task 8.1   Introduction stages

Read the Introduction of your selected Provided Example Article (PEA), decide if all stages are present, and mark where each one begins and ends. (Remember that it is possible that stages may be repeated or come in a different order to that suggested in Figure 8.1.)

Compare your findings with our suggestions in the Answer pages.

Now, do the same for your own Selected Article (SA). Discuss your findings with a colleague or teacher, if appropriate.

8.2 Stage 1: Locating your project within an existing field of scientific research

Constructing the right setting for your paper

In Stage 1, authors mostly begin with broad statements that would generally be accepted as fact by the members of their reading audience. The present tense is often used for this kind of statement because one function of the present tense in English is expressing information perceived as always true. Sentences written in the present perfect tense are also common in Stage 1, expressing what has been found over an extended period in the past and up to the present. These statements may or may not include references, depending on the field and the topic of the paper. Now do Task 8.2.

Task 8.2   Introduction Stage 1 analysis

1 Check the first paragraphs of the Introductions of the three PEAs and complete Table 8.2. Then check your answers with our suggestions in the Answer pages.
2 Repeat the exercise for your SA, compare your findings with those for the PEAs, and discuss any differences with a colleague or teacher, if appropriate.

Table 8.2   Introduction Stage 1 analysis.

Question	Kaiser et al. (2003)	Britton-Simmons & Abbott (2008)	Ganci et al. (2012)
Are some sentences written in the present tense? How many?			
Are some sentences written in the present perfect tense? How many?			
Which tense is used more? Why do you think this is the case?			
How many sentences contain references?			
What kinds of sentences do not have references?			

Authors next seek to move their readers smoothly from these broad, general statements towards one sub-area of the field, and then to the authors' own particular topic. One way to think about this is to imagine you are looking at a selected "country" on a map (the broad area where the Introduction begins), then zooming in on a particular "province," and finally focusing on a particular "city" (the topic area of research to be presented in the paper). Work through Task 8.3 to help you understand this way of thinking about Stage 1 in Introductions.

Writers move their readers through these steps by linking their sentences through the positioning of *old* and *new information*. Old information is any information that the reader already knows; it is placed towards the beginning of sentences. New information comes towards the end of sentences. (This convention is very important for improving flow in all forms of technical writing.) See Task 8.4.

Task 8.3 "Country" to "city" in Stage 1

1 Look at the Introduction of your selected PEA. What is the "country"? The "province"? The "city"?

Check your answers against our suggestions in the Answer pages.

2 Now do the same task for the Introduction to the SA you are analysing. "Country"? "Province"? "City"?

3 Now try to suggest these three features for your Own Article (OA). Remember, your "city" is not your purpose for conducting the study, but rather the specific topic area for your paper. "Country"? "Province"? "City"?

Task 8.4 Identifying old or given information

Look at the following extract from the Introduction in Kaiser et al. (2003) and underline the words that represent or refer to *old information* (information the reader already knows about, also called *given information*):

Legumes form symbiotic associations with N_2-fixing soil-borne bacteria of the *Rhizobium* family. The symbiosis begins when compatible bacteria invade legume root hairs, signalling the division of inner cortical root cells and the formation of a nodule. Invading bacteria migrate to the developing nodule by way of an "infection thread," comprised of an invaginated cell wall. In the inner cortex, bacteria are released into the cell cytosol, enveloped in a modified plasma membrane (the peribacteroid membrane (PBM)), to form an organelle-like structure called the symbiosome, which consists of bacteroid, PBM and the intervening peribacteroid space (PBS; Whitehead and Day, 1997). The bacteria, subsequently, differentiate into the N_2-fixing bacteroid form. The symbiosis allows the access of legumes to atmospheric N_2, which is reduced to NH_4^+ by the bacteroid enzyme nitrogenase. In exchange for reduced N, the plant provides carbon to the nodules to support bacterial respiration, a low-oxygen environment in the nodule suitable for bacteroid nitrogenase activity, and all the essential nutritional elements necessary for bacteroid activity. Consequently, nutrient transport across the PBM is an important control mechanism in the promotion and regulation of the symbiosis.

Check your answers in the Answer pages.

8.3 Using references in Stages 2 and 3

In Stages 2 and 3 of an Introduction (see Figure 8.1), authors use selected literature from their field to justify their study and construct a gap, niche, or need for extension for their own work. They write sentences supported by *references* to the literature they have selected. In this context, the term *literature* refers to all the published research articles, review articles, and books in a given field. The term also includes information published on websites that have been peer-reviewed or belong to organisations with appropriate scientific reputations.

References to other published studies, also known as citations or in-text citations, can be used in all stages of the Introduction, as you have seen in the samples we have looked at. They appear in the text either as a surname and year in brackets, e.g. (McNeill 2000), or as a number, e.g. [7]. The details of the presentation depend on the style stipulated by the journal. Check the Instructions to Contributors of your target journal for the necessary information on referencing style. These references *refer* to the list of references at the end of the paper, where the full publication details are written, again using the style stipulated by the target journal.

Citations are particularly vital in showing that you know clearly the work that has been conducted by others in your "city" area (see Task 8.3), and therefore what has not been done and needs to be done: the gap that your study will help to fill. This function is carried out in Stages 2 and 3. What you are required to do here is, in effect, to construct an argument which justifies your own study and shows why and how it is important.

Using citation to develop your own argument

This section provides examples of parts of paragraphs using three different citation methods (the references cited have been invented for demonstration purposes only). These methods can be called (1) information prominent, where the focus of the sentence is only on the information being presented; (2) author prominent, where the name of the author of the information is given prominence in the sentence; (3) and weak author prominent, where the concept of author is given prominence, but author names do not appear in the main part of the sentence. Observe how the different methods contribute to the way in which the writer's argument is developed. (N.B. For this section, the term *author* is used for the author of a published paper that is being cited; the term *writer* is used to refer to the person writing the text that cites the author's work.)

Information prominent citation
Shrinking markets are also evident in other areas.[1] The wool industry is experiencing difficulties related to falling demand worldwide since the development of high-quality synthetic fibres (Smith 2000).

This is the default style in many areas of science and is the only style used in the Introductions of two of the PEAs. However, the other two options should also be part of a writer's repertoire, for use when appropriate.

Author prominent citation style 1
Shrinking markets are also evident in other areas. As Smith (2000) pointed out, the wool industry is experiencing difficulties related to falling demand worldwide since the development of high-quality synthetic fibres.

[1] This first sentence is a "topic sentence" for the paragraph: its function here is to form a link to the previous paragraph (which discussed shrinking markets), and to alert the reader to the topic of the current paragraph. Topic sentences are an effective way of creating logical flow in science writing.

This style gives more option to show the writer's view of the cited fact. In this case, it shows that the writer (you!) agrees with Smith.

Author prominent citation style 2

Shrinking markets are also evident in other areas. Smith (2000) argued that the wool industry was experiencing difficulties related to falling demand worldwide since the development of high-quality synthetic fibres. However, Jones et al. (2004) found that industry difficulties were more related to quality of supply than to demand issues. It is clear that considerable disagreement exists about the underlying sources of these problems.

This style also allows the use of verbs such as *argued*, which gives the reader advance notice that a *however* or some other contrast may be coming, and indicates that what is being cited is not necessarily accepted as correct by you, the writer. However, there is a danger attached to the author prominent style. If it is over-used, it can make the text sound like a list, rather than a logically constructed argument. We recommend that you use this style sparingly, perhaps when you are approaching the specifics of the gap your study will address. One example (of two) from the Introduction of the PEA by Ganci et al. (2012) demonstrates this feature:

> These modeling efforts all used infrared satellite data acquired by AVHRR and/or MODIS to estimate the time-averaged discharge rate (TADR) following the methodology of Harris et al. (1997). Recently, we developed . . .

It is also useful to pay close attention to the papers you read in your own field, to check how often, if at all, this style appears.

Weak author prominent citation

Several authors have reported that the wool industry is experiencing difficulties related to falling demand since the development of high-quality synthetic fibres (Smith 2000, Wilson 2003, Nguyen 2005). For example, Smith (2000) highlighted . . .

This method has a general reference to authors in the subject and then more than one reference in brackets. It is followed here by an author prominent citation. This style can be useful as a topic sentence when beginning a new subtopic or line of argument. Note that this style requires the use of the present perfect tense (*have reported*).

Writers choose their citation method to fit with the way their paragraph advances their argument (see Table 8.3).

Citing when you cannot obtain the original reference

Editors usually require that writers cite only those papers that they have actually read. However, if you cannot obtain an original article and are therefore obliged to rely on another author's interpretation of a fact or finding you want to cite, you may use the following form of secondary citation in-text:

> [The finding or fact you want to cite] (Smith 1962, cited in Jones 2002).

In such cases, only Jones (2002) appears in the reference list.

Table 8.3 Use of different citation styles in a segment of the Introduction from McNeill et al. (1997). Text reproduced by permission of CSIRO Publishing.

Introduction text	Citation style
Foliar feeding does not disturb the system and has the additional advantage that shoots tolerate higher concentrations of N than roots (Wittwer et al. 1963).	Information prominent
Spray application of ^{15}N-labelled urea has been successfully used to label legumes *in situ* under field conditions (Zebarth et al. 1991) but runoff of ^{15}N-labelled solutions from foliage to the soil will complicate interpretation of root-soil dynamics.	Information prominent
	Writer's evaluation statement
Russell and Fillery (1996), using a stem-feeding technique, have shown that in situ ^{15}N-labelling of lupin plants growing in soil cores enabled total below-ground N to be estimated under relatively undisturbed conditions, but they indicated that the technique was not adaptable to all plants, particularly pasture species. Feeding of individual leaves with a solution containing ^{15}N is a technique that has been widely used for physiological studies in wheat (Palta et al. 1991) and legumes (Oghoghorie and Pate 1972; Pate 1973).	Author prominent
	Information prominent
The potential of the technique for investigating soil-plant N dynamics was noted as long as 10 years ago by Ledgard et al. (1985) following the use of ^{15}N leaf-feeding in a study of N transfer from legume to associated grass.	Author prominent, but using the passive voice so that the link (technique) can come first in the sentence as old information

Source: McNeill, A.M., Zhu, C.Y., & Fillery, I.R.P., Use of in situ 15N-labelling to estimate the total below-ground nitrogen of pasture legumes in intact soil-plant systems. *Australian Journal of Agricultural Research*, 48, 295–304. © 1997 CSIRO Publishing.

8.4 Avoiding plagiarism when using others' work

Another important reason to pay careful attention to referencing is to avoid plagiarising other people's work unintentionally. Plagiarism is using data, ideas, or extended wordings that originated in work by another person without appropriately acknowledging their source. It is generally regarded as a form of cheating in academic and publishing contexts, and papers will be rejected if plagiarism is detected. Incomplete citation also prevents your gaining credit for knowing the work of other researchers in the field. Effective and inclusive citation helps you present yourself as a knowledgeable member of the research community, which can be important in terms of the impression you make on referees evaluating your manuscripts. It also allows others to benefit from the sources of information you have used.

Avoiding inadvertent plagiarism requires writers to do two things: to be aware of the kinds of situations where such plagiarism is likely to occur, and to develop effective note-taking practices to ensure they remain aware of the status of their notes as they convert them into sentences in a paper for submission. The important thing to watch for is that it is clear to your reader whether the idea or fact you are using in each and every sentence is your own or has come from the work of another person. If it comes from someone else's work, cite them! It is possible that the person whose idea it originally was will be a referee of your paper, and they will be sure to notice

> ### Task 8.5   Identifying plagiarism
>
> Presented here are two versions of the same information, adapted from the Introduction in McNeill et al. (1997). In Version 2, identify where the rewrite causes the writers to plagiarise by writing in their own voice ideas that originated in another document (as demonstrated in Version 1).
>
> **Version 1** Russell and Fillery (1996), using a stem-feeding technique, have shown that *in situ* ^{15}N-labelling of lupin plants growing in soil cores enabled total below-ground N to be estimated under relatively undisturbed conditions, but they indicated that the technique was not adaptable to all plants, particularly pasture species.
>
> **Version 2** Russell and Fillery (1996), using a stem-feeding technique, have shown that *in situ* ^{15}N-labelling of lupin plants growing in soil cores enabled total below-ground N to be estimated under relatively undisturbed conditions. However, this technique is not adaptable to all plants, particularly pasture species.
>
> Check your answers in the Answer pages.

the problem. In any case, the referees will know the literature well, so it is very important to be accurate in your citation practices. Now complete Task 8.5.

Remember also that direct quotations using quotation marks or inverted commas ("...") are extremely rare in science writing. This means that authors need to paraphrase sentences that appear in the work of other authors, rather than copying them verbatim. However, remember also that you can expand your repertoire of sentence structures by removing the content (most often the noun phrases, indicated by NP in the following example) from sentences that appeal to you and re-using the shell (or sentence template) for your own content. For example, from the sentence in Task 8.5, Version 1, you could reuse this shell:

[Authors], using [NP1], have shown that [NP2] enabled [NP3] to be estimated under [adjective] conditions, but they indicated that the technique was not adaptable to all [NP4], particularly [NP5].

See Chapter 17 for more details on this approach.

8.5   Stage 3: Indicating the gap or research niche

This can be done in a multitude of ways. As discussed previously, authors often present a broad gap early in the Introduction, and a more specific one close to the end. Examples include the following, taken from the PEA by Britton-Simmons & Abbott (2008):

> However, understanding how these processes interact to regulate invasions remains a major challenge in ecology.
>
> Despite its acknowledged importance, propagule pressure has rarely been manipulated experimentally and the interaction of propagule pressure with other processes that regulate invasion success is not well understood.
>
> It is presently unclear how different disturbance agents influence long-term patterns of invasion.

It is common to find so-called signal words that indicate that a Stage 3 statement is being made. In the examples from Britton-Simmons & Abbott (2008), such signal words include *however*, *remains a major challenge*, *rarely*, *not well understood*, and *presently unclear*. Now complete Tasks 8.6 and 8.7.

Task 8.6   Signal words for the research gap or niche

Reread the Introductions from McNeill et al. (1997) (see Table 8.1) and your selected PEA, and identify the signal words that indicate a gap is being described. List them and then check the list against our suggestions in the Answer pages.

Task 8.7   Drafting your own Introduction: Stage 3

Begin to draft Stage 3 for the Introduction of your own paper, if appropriate.

8.6 Stage 4: The statement of purpose or main activity

At the end of the Introduction, authors set up their readers' expectations for the rest of the paper: they tell them what they can expect to learn about the research being presented. As indicated in Figure 8.1, Stage 4 of the Introduction is generally in the form of the aim or purpose of the study to be reported, or its principal activity or finding. Which of these appears in a particular journal depends on the conventions of the research field and of the journal itself. Within these constraints, authors have considerable flexibility in choosing how they will word their Stage 4, and it can be instructive to pay attention to how this is done in each paper that you read for your research. You may like to keep a list of possible wordings, to help when you come to the writing of your own papers (see Task 8.8). Now complete Task 8.9.

Task 8.8   Stage 4 sentence templates

Identify the Stage 4 in the Introduction in McNeill et al. (1997), presented in Table 8.1, and in your selected PEA. We have provided a shell, or sentence template, from each one in the Answer pages.

Task 8.9   Drafting your own Introduction: Stage 4

Draft a Stage 4 for the Introduction of your own paper, if appropriate. Write it so that it runs smoothly on from your Stage 3 gap statement, to form the closing part of your Introduction. Make sure that all the keywords in your title (see Chapter 10) have been used in these sentences, to meet the expectations you set up for your readers when they read the title.

8.7 Stages 5 and 6: Highlighting benefit and mapping the article

Neither of these stages is obligatory in a given Introduction; for example, only two of the three PEAs include a Stage 5 (highlighting a value or benefit of the study reported) and none of them contains a Stage 6 (an explicit "map" or description of how the rest of the article is arranged). If present, Stage 5 can appear anywhere in the Introduction – remember, the stage numbering is only a naming device, not an immutable indication of the order of the sections. Nevertheless, the two Stage 5s in the current discussion (in Britton-Simmons & Abbot 2008 and Ganci et al. 2012) do occur at the end of their respective Introductions, after the Stage 4s. Both serve the purpose of emphasising the significance of the study design or outcomes for the field.

The need for a Stage 6 in a research article Introduction depends on the conventions of the research field and the journal; for example, this stage is commonly found in articles in physics and computer science. Once again, check your target journal to make sure of what is expected in your own case.

8.8 Suggested process for drafting an Introduction

Here is a summary of a process for drafting an Introduction. It is useful *after* you have made the key decisions about the results you will include in the paper, and what they mean for the audience who will read it.

1   Begin with Stage 4. Write an aim statement, or a statement describing what the paper sets out to do or show. This is usually the easiest part of the Introduction to write. It will appear in the final paragraph of the Introduction, but it is useful to write it early in the drafting process. Make sure that this statement includes all the parameters or issues that make your study new and significant, so that readers are not surprised by meeting important aspects of your study later on in the document that they were not prepared for in your Stage 4.

2   Draft Stage 3 next: the gap or need for further work. As we have seen in the previous sections, there may be one or more sub-gaps at different places in your Introduction, as well as a Stage 3 statement that leads into Stage 4. Consider beginning your Stage 3 sentences with words such as *however* or *although*, and incorporating words indicating a need for more research, such as *little information*, *few studies*, *unclear*, or *needs further investigation*.

3   Then think about how to begin Stage 1, the setting. Think about your intended audience and their interests and background knowledge, and the ideas you have highlighted in your title. Try to begin with words and concepts that will immediately grab the attention of your intended readers.

4   Next, arrange the information you have collected from the literature into Stage 2, or a series of Stage 2/3 segments that justify the inclusion of different parts of your Stage 4. This is a very important part of the Introduction, and you will probably need quite a bit of time to write it. You may need to do some more searching of the literature to make sure you have done the best possible job of finding the relevant work in the area and the most recent studies.

5 Combine the stages into a coherent Introduction. You may need to add additional sentences providing background, and/or rearrange sentences or parts to get the best possible logical development. Section 8.9 focuses on strategies for revising your Introduction to enhance the logical flow of the writing, once you are happy with the content you have included.

8.9 Editing for logical flow

In English writing, the responsibility rests with the writer to ensure that the reader recognises the logical flow of the argument being presented. This is not the case in all languages! However, even for writers with English as a first language (EL1), the strategies for achieving this goal in their writing are often not obvious. We suggest some important ones in the following subsections. We have mentioned several of these previously in the book, but this section brings them together into a coherent set and provides you with some practice in improving poor examples.

Strategy 1: Always introduce ideas

Use informative titles, subheadings, and introduction sections to set up expectations in your reader. A key to effective scientific and technical communication in English is to set up expectations in the reader's mind, and then meet those expectations as soon as possible.

Make the wording of your subheadings – if your target journal uses them – a part of the process of telling your reader what to expect next, in much the same way that the paper's title alerts them to the main message of the paper as a whole. In paragraphs, use the first sentence as a topic sentence to orient your readers to the main point or purpose of the paragraph. Topic sentences can also be used to link the upcoming paragraph to the one that precedes it; see Task 8.10.

Task 8.10 Topic sentence analysis

What information would you expect to find in the paragraph introduced by each of the following sentences? What do you think was the focus at the end of the previous paragraph?

1 Two classes of putative Fe(II)-transport proteins (Irt/Zip and Dmt/Nramp) have been identified in plants (Belouchi et al., 1997; Curie et al., 2000; Eide et al., 1996; Thomine et al., 2000).
2 Propagule pressure is widely recognized as an important factor that influences invasion success (MacDonald et al. 1989; Simberloff 1989; Williamson 1996; Lonsdale 1999; Cassey et al. 2005).
3 On Etna, the MAGFLOW Cellular Automata model has successfully been used to reproduce lava flow paths during the 2001, 2004 and 2006 effusive eruptions (Del Negro et al., 2008; Herault et al., 2009; Vicari et al., 2007).

Check these paragraphs in the PEAs by Kaiser et al. (2003), Britton-Simmons & Abbott (2008), and Ganci et al. (2012) to find out if your predictions are correct, and see our comments in the Answer pages.

> Now look at an article you have not read before and read the first sentences of each of the paragraphs in the Introduction. Can you predict the content of the paragraphs? N.B. The first sentence is very often, but not always, the topic sentence of the paragraph.

Strategy 2: Move from general information to more specific information

Readers of English text expect that they will read general information about any topic or point first, before encountering details, examples, or other more specific information.

Consider the following sample paragraph and decide whether it meets the requirement to move from the general to the particular. Alternatively, is there a sentence that seems to be too general late in the paragraph? (Sentences are numbered to make it easier to refer to them later.)

> [1]Pleuropneumonia (APP) can present as a dramatic clinical disease or as a chronic, production limiting disease in pig herds. [2]A sudden increase in the number of sick and coughing pigs and a sharp rise in mortalities among grower/finisher pigs may herald an outbreak of APP in a herd. [3]On the other hand, signs may be limited to a drop in growth rate and an increase in grade two pleurisy lesions in slaughter pigs. [4]The disease surfaced in the Australian pig population during the first half of the 1980s and ten years later was regarded as one of the most costly and devastating diseases affecting the Australian pig industry.

Do you agree that Sentence 4 is more *general* than the other sentences? In that case, the paragraph could be improved by moving Sentence 4 to the beginning of the paragraph, as follows. Some slight changes of wording have also been made to improve the sense.

> Pleuropneumonia (APP) surfaced in the Australian pig population during the first half of the 1980s and ten years later was regarded as one of the most costly and devastating diseases affecting the Australian pig industry. It can present as a dramatic clinical disease or as a chronic, production limiting disease in pig herds. A sudden increase in the number of sick and coughing pigs and a sharp rise in mortalities among grower/finisher pigs may herald an outbreak of APP in a herd. On the other hand, signs may be limited to a drop in growth rate and an increase in grade two pleurisy lesions in slaughter pigs.

Strategy 3: Put old (or given) information before new information

To understand the basis of this recommendation, consider first the following two short paragraphs. Both contain exactly the same information, but in a different order: decide whether one version is easier to understand than the other.

> **Version A** [1]Clay particles have surface areas which are many orders of magnitude greater than silt or sand sized particles. [2]The ability of soils to shrink when dried is controlled by the interactions of these clay surfaces with water and exchangeable cations.
>
> **Version B** [1]Clay particles have surface areas which are many orders of magnitude greater than silt or sand sized particles. [2]The interactions of these clay surfaces with water and exchangeable cations control the ability of soils to shrink when dried.

Readers usually agree that Version B is easier to follow. The following paragraph seeks to explain why this should be so. When readers begin to read Sentence 2 of either version, they already know all the information that is included in Sentence 1; therefore, all the Sentence 1 information can be described as old or given information in this context. In Version A, it is not until the second half of Sentence 2 that readers encounter a reference to this old information again (clay surfaces). All the information at the beginning of Sentence 2 is new information, and so the sentence does not follow the recommendation to put old information before new information. This structuring contributes to making the passage difficult to follow. In Version B, the information order has been changed to put the old information at the beginning of Sentence 2 and the new information at the end. Now complete Task 8.11 to apply this "old before new" idea.

Task 8.11   Old information before new information

Which sentence in the following paragraph needs to be changed in order to follow Strategy 3?

Pleuropneumonia (APP) surfaced in the Australian pig population during the first half of the 1980s and ten years later was regarded as one of the most costly and devastating diseases affecting the Australian pig industry. It can present as a dramatic clinical disease or as a chronic, production limiting disease in pig herds. A sudden increase in the number of sick and coughing pigs and a sharp rise in mortalities among grower/finisher pigs may herald an outbreak of APP in a herd. On the other hand, signs may be limited to a drop in growth rate and an increase in grade two pleurisy lesions in slaughter pigs.

Check your answer in the Answer pages.

Strategy 4: Make a link between sentences within the first seven to nine words

Another way to describe the difference between Versions A and B under Strategy 3 relates to how many words the reader has to read in the next sentence (Sentence 2 in each version) before encountering a link with what is already known (the old information). In Version A, the reader has to read 15 words before finding the first link, which is the word *clay*. In Version B, however, the first link word comes as word five of Sentence 2. Making this link within the first seven to nine words of a sentence enhances readability: that is, the ease with which readers will process the information presented. Sentence 3 in Task 8.11 works better when it is rewritten as follows:

An outbreak of APP in a herd may be heralded by a sudden increase in the number of sick and coughing pigs and a sharp rise in mortalities among grower/finisher pigs.

In this version, the fourth word (*APP*) provides the old information, and old information precedes new information. The method used to change the information order in the sentence was to change an active voice verb, *may herald*, to a passive voice verb, *may be heralded*. This method is often useful to improve flow within paragraphs. In our opinion, promoting flow in this way is a more important consideration than avoiding the passive voice at all costs, as is sometimes recommended in writing manuals.

Read the following two sentences and consider how easy they are to follow.

> [1]The definition of seed quality is very broad and encompasses different components for different people. [2]The quality and quantity of flour protein, dough mixing requirements and tolerance, dough handling properties and loaf volume potential are quality parameters of wheat seed for bread bakers.

Sentence 2 is not easy to follow because readers have to read a very long subject of 19 words before they arrive at the verb *are*. Sentences with very long subjects and short verbs at the end are often called top-heavy sentences. In both of the following edited versions, Sentence 2 has been changed so that the verb and its subject fit within the first seven to nine words, and the list of items (which makes up the new information in the sentence) comes at the end:

> **Edited version A** [1]The definition of seed quality is very broad and encompasses different components for different people. [2]Quality parameters of wheat seed for bread bakers are the quality and quantity of flour protein, dough mixing requirements and tolerance, dough handling properties and loaf volume potential.

> **Edited version B** [1]The definition of seed quality is very broad and encompasses different components for different people. [2]For bread bakers, quality parameters of wheat seed are the quality and quantity of flour protein, dough mixing requirements and tolerance, dough handling properties and loaf volume potential.

As a general rule, if you want to write a list, it should come at the end of its sentence. Now complete Tasks 8.12 and 8.13.

Task 8.12   Revising top-heavy sentences

Change these top-heavy sentences so that each has a verb and its subject within the first seven to nine words:

1 In this project the *Rhizoctonia* populations of two field soils in the Adelaide Plains region of South Australia were characterised.
2 A balance between deep and shallow rooting plants, heavy and light feeders, nitrogen fixers and consumers and an undisturbed phase is needed to achieve maximum benefit through rotation.

Compare your answers with the suggested improvements in the Answer pages.

Task 8.13   Revising your own Introduction for flow

If you are writing a draft Introduction, take the time now to revise it using the strategies discussed in this chapter.

CHAPTER 9

The Discussion section

The Discussion section plays an important part in demonstrating to the audience of editor and reviewers that the contribution is new and significant. It does this by contextualising the main research findings within accepted knowledge, with emphasis on qualifying what is new and important by reference to any limitations and by speculation about the greater implications of the new knowledge.

A good starting point for discussion writing is to return to the bullet points you prepared when you were working on Chapter 4. Think of what you did then as "mapping" your article's contribution, and use the bullet points labelled "D" (for discussion) as prompts to help you begin drafting this section.

9.1 Important structural issues

There are several important issues to think about as you begin to draft your Discussion section, or text that discusses results under other headings. (Although this chapter is framed around separate Discussion sections, the guidelines can be applied to discussion text wherever it appears.)

Structure of the Discussion

- Does the journal you are targeting allow the option of a combined Results/Discussion section (or other combinations of content under self-chosen headings), followed by a separate Conclusion? Would this arrangement suit your story?
- Does the journal permit a separate Conclusion section where the Discussion section is relatively long? Would your paper benefit from one?
- Does the journal publish Discussion sections which include subheadings? Would this option help you signal your main messages to the reader? (Do Task 9.1 now to further clarify this issue.)

Relating the Discussion closely to the paper's title

- As you decide on the key elements of the paper's story that will be emphasised in the Discussion, consider redrafting the title to reflect them more clearly.

Writing Scientific Research Articles: Strategy and Steps, Third Edition.
Margaret Cargill and Patrick O'Connor.
© 2021 John Wiley & Sons Ltd. Published 2021 by John Wiley & Sons Ltd.

Task 9.1 Structures of Discussion sections

Check the location of Discussion text in your selected Provided Example Article (PEA).

- Is there a single separate section called Discussion?
- Does it include subheadings?
- Does discussion text appear under other subheadings?
- Does the paper also have a separate section headed Conclusion(s) or similar?

Now answer the same questions about your Selected Article (SA). Discuss your findings with a colleague or teacher, if appropriate. Why do you think the authors chose the arrangement they did? Do you think the Discussion could have been improved by using a different arrangement?

Relating the Discussion closely to the Introduction

- Remember that you need to ensure that your Discussion connects clearly with the issues you raised in your Introduction, especially the "country" where you began (see Section 8.2), the evidence leading up to your Stage 3 gap or research niche, and your statement of purpose or main activity. When the first draft of the Discussion is ready, go back to the Introduction and check for a close fit. If necessary, redraft the Introduction to make sure the issues of importance in the Discussion appear there also.
- However, it is not necessary to include in the Introduction all the literature that will be referred to in the Discussion. It is important not to repeat information unnecessarily in the two sections.

9.2 Information elements to highlight the key messages

The types of information commonly included in Discussion sections are as follows – this list can form a checklist for you as you write. You may not have something to say under every point in the list for every result you discuss, but it is worthwhile thinking about each element in turn as you draft the section.

1 A reference to the main purpose or hypothesis of the study, or a summary of the main activity of the study.
2 A restatement or review of the most important findings, generally in order of their significance, including

   a. whether they support the original hypothesis, or how they contribute to the main activity of the study, to answering the research questions, or to meeting the research objectives; and
   b. whether they agree or contrast with the findings of other researchers.

3 Explanations for the findings, supported by references to relevant literature, and/or speculations about the findings, also supported by literature citation.
4 Limitations of the study that restrict the extent to which the findings can be generalised beyond the study conditions.

5 Implications of the study (generalisations from the results: what the results mean in the context of the broader field).
6 Recommendations for future research and/or practical applications (after Weissberg & Buker 1990).

The elements numbered 2–5 are often repeated for each group of results that is discussed.

When you are analysing Discussion sentences in papers in your field, pay special attention to how they manage Point 4: limitations. Generally, limitations restrict the extent to which findings can be generalised or applied to situations beyond those described in the paper. However, if authors write about these restrictions as throwing up a block or barrier to further understanding, they can claim that uncovering them is an additional contribution made by the paper.

Structuring your writing is important when drafting this section; think about the main points you want your reader to understand from the Discussion, and consider using subheadings or topic sentences to highlight where it focuses on each of these points. This strategy can help you write text suitable to be used by future readers as the basis of citations to your work.

Now complete Tasks 9.2, 9.3, and 9.4.

Task 9.2   Information elements in the Discussion section

Select the part of this task that relates to your selected PEA:

1 From Kaiser et al. (2003), read the second subsection of the Discussion, under the heading "Specificity of GmDmt1;1." For each sentence, based on the checklist provided in the text, identify the information element(s) that are presented.
2 From Britton-Simmons & Abbott (2008), read the first paragraph of the Discussion. For each sentence, based on the checklist provided in the text, identify the information element(s) that are presented.
3 From Ganci et al. (2012), read the first paragraph of "5.3 Discussion." For each sentence, based on the checklist provided in the text, identify the information element(s) that are presented.

Check your answers in the Answer pages.

Task 9.3   Analysing a Discussion section

Select one or more paragraphs from the Discussion section of your SA to use for a similar analysis to the one you performed for Task 9.2.

• For each sentence, identify the information element(s) presented.
• Can you identify any strategies the authors have used to clarify the key messages of their Discussion section (subheadings, topic sentences)?
• Is there a close link between the key or "take-home" messages and the paper title?

Discuss your findings with a colleague or teacher, if appropriate.

Task 9.4   Drafting your own Discussion section

Begin to draft the Discussion section of your own paper, if appropriate, referring to your "map" from Chapter 4 and using the checklist in the text to ensure you include all the relevant information elements.

9.3 Negotiating the strength of claims

For the last four information elements mentioned in Section 9.2, authors need to pay particular attention to the *verbs* they use to comment on their results. They carry much of the meaning about *attitude to findings* and *strength of claim*.

To illustrate how this works, we use sentences containing *that*. In these sentences, authors have two opportunities to show how strong they want their claim to be:

- in the choice of vocabulary and tense in the main verb; and
- in the choice of verb tense in the *that* clause.

Let us look at some example sentences from the PEAs. The verb phrases of interest are underlined in the tabular presentations in Table 9.1.

In Example 1, the main verb is in the present tense (indicating that it is "always true," a very strong statement) and the meaning of the verb itself (*demonstrate*) is also strong; the verb in the *that* clause is also in the present tense. Together, these choices indicate that the authors are very confident of the claim they make in this sentence. That is, they think that the data they have presented in the article are strong enough to justify making the strongest possible statement about what the results mean.

Example 2 is of similar strength to Example 1: *means* is similar in strength of certainty to *demonstrate*, and the present tense is used in both the main clause and the *that* clause.

Example 3 is not as strong overall as the first two examples, because although the present tense is used in the main clause, *indicate* shows less strength of certainty than *demonstrate*; the verb in the *that* clause is in the future tense, indicating a strong prediction of outcome.

In Example 4, a much weaker verb is used in the main clause: *appears* (which is only ever used with the subject *it* in this kind of sentence). The verb in the *that* clause is in the present tense, reflecting the strength of the evidence the authors have presented earlier in the paragraph.

In Example 5, the main clause verb *suggests* is again weak in terms of its level of certainty; in addition, the verb in the *that* clause has been made less definite by the use of the modal verb *may*. Thus, Example 5 makes the weakest claim of any of the sentences we have considered here. This is not a bad thing at all: the important thing for authors is that they match the strength of their sentences (using the vocabulary and tense options just discussed) with the strength of the data and arguments they have presented in the Results and Discussion sections of their paper. This is a key feature that is checked by referees during review of a manuscript, and by thesis examiners as well.

Now complete Task 9.5 to consolidate your learning.

Table 9.1 Five examples of language choice, vocabulary, tense, and modality in Discussion text.

Example number	Subject of main verb	Main verb	"That" plus subject of "that" clause	Verb from "that" clause	Rest of sentence
1	Our experimental results	demonstrate	that space- and propagule-limitation both	regulate	*S. muticum* recruitment.
2	This	means	that running MAGFLOW on GPUs	provides	a simulation spanning several days of eruption in a few minutes.
3	These results	indicate	that *S. muticum* recruitment under natural field conditions	will be determined	by the interaction between disturbance and propagule input.
4	. . . it	appears	that GmDmt1;1	has	the capacity to function *in vivo* as either an uptake or an efflux mechanism in symbiosomes.
5	The presence of an IRE motif	suggests	that GmDmt1;1 mRNA	may be stabilized	by the binding of IRPs in soybean nodules when free iron levels are low.

Task 9.5   Negotiating strength of claims with verbs

Complete the schematic in Table 9.2 by listing alternative choices for the underlined words, writing them in increasing order of strength down the page. The strongest alternatives have been completed as an example.

 Check your answers with our suggestions in the Answer pages.

Table 9.2 Negotiating strength of claims with verbs – an exercise in ranking possible verb forms in a Discussion sentence in order of strength of claim.

The presence of an IRE motif	suggests ↓ demonstrates	that GmDmt1;1 mRNA	may be stabilized ↓ is stabilized	by the binding of IRPs in soybean nodules when free iron levels are low.	Weak ↓ Strong

An alternative construction without a *that* clause is also found in science writing. Look at the following example, taken from the PEA by Britton-Simmons & Abbott (2008):

> Previous studies have demonstrated a positive relationship between propagule pressure and the establishment success of non-native species.

In this construction, the object of the verb is a noun phrase, here "a positive relationship between propagule pressure and the establishment success of non-native species." It is interesting to note that when this construction is used, the author does not need to make a decision about what tense to use in the *that* clause.

The effects of the different language choices discussed in this section are also highly relevant when you write sentences about findings reported in the literature, either in Discussion sections or in Introductions. For example, consider this sentence from the Discussion of Britton-Simmons & Abbott (2008):

> A more recent terrestrial experiment also reported that propagule pressure was the primary determinant of invasion success, overwhelming the effects of other factors, such as disturbance and resident diversity, which were concurrently manipulated (Von Holle & Simberloff 2005). However, . . .

Both the main verb (*reported*) and the verb in the *that* clause (*was*) are in the past tense, indicating that the point the authors wish to make is that the citation refers to a situation limited to the conditions of the cited study only, in the past. The following *however* introduces a further explanation of the limitation. This ability to choose your tense is a powerful tool in writing argument involving the use of citations.

Now complete Task 9.6.

Task 9.6   Analysing and practising strength of claim

Reread the Discussion section of your selected PEA and find sentences that use both patterns (with and without a *that* clause) in the Discussion or Conclusion sections. Identify the verbs that carry the strength-of-claim messages, and discuss your findings with a colleague or teacher, if appropriate.

Then consider your own results and begin to draft sentences to comment on them in your Discussion section, paying particular attention to matching the strength of the claim in your sentences to the strength of your data and arguments.

CHAPTER 10

The title and keywords

The title and keywords you finally select for your manuscript form an important part of your communication with your readers – both the editor and referees who will evaluate the paper and the members of your discipline community who you want to read it after its publication. From the reviewer criteria we considered in Chapter 3, we know it is important that the title clearly indicates the content of the paper, but there are various ways in which this can be achieved. In this chapter, we look at advice about attracting the attention of your target readers effectively.

10.1 Strategy 1: Provide as much relevant information as possible, but be concise

The purpose of a title is to attract busy readers in your particular target audience, so that they will want to access (download) and read the abstract, and potentially the whole article. The more revealing your title, the more easily your potential readers can judge how relevant your paper is to their interests. To exemplify the importance of this issue, we quote from relevant Author Guidelines.

The *Journal of Ecology* asks for "a concise and informative title" (https://besjournals.onlinelibrary.wiley.com/hub/journal/13652745/author-guidelines). "Concise" is a difficult term to interpret: general guidelines suggest 50–140 characters, and a recent study of 274 clinical medical titles found a mean of 16.6 words (Kerans et al. 2020), up from 15 words in a similar earlier study (Kerans et al. 2016). The *New Phytologist* stipulates a concise and informative title (for research papers, ideally stating the key finding or framing a question) (https://nph.onlinelibrary.wiley.com/hub/journal/14698137/about/author-guidelines). We will return to the question of the most effective grammatical form for titles later.

10.2 Strategy 2: Use carefully chosen keywords prominently

It is important to take care in deciding which words (keywords) will capture the attention of readers likely to be interested in your paper. You can tackle this decision from two directions: top-down – words colleagues could use to search for your paper; and bottom-up – words you have used frequently in your paper. Once

Writing Scientific Research Articles: Strategy and Steps, Third Edition.
Margaret Cargill and Patrick O'Connor.
© 2021 John Wiley & Sons Ltd. Published 2021 by John Wiley & Sons Ltd.

chosen, place your keywords near the front of your title. This practice also helps ensure that your title is picked up efficiently by literature-scanning services and search engines, which use a keywords system to identify papers of interest to particular audiences and rank them in order of relevance to searches entered online. Wherever possible, it is a good idea to place the most important word(s) in your title in the position of power: the beginning. For example:

✗ Effects of added calcium on salinity tolerance of tomato
✓ Calcium addition improves salinity tolerance of tomato

One effective way to ensure your keyword(s) are at the front of your title is to use a colon (:) or a dash (–) to separate the first, keyword-containing part of the title from a second, explanatory section. Effective examples include the following (taken from the reference lists of the Provided Example Articles, PEAs):

✓ Disturbance, invasion, and reinvasion: managing the weed-shaped hole in disturbed ecosystems
✓ Native weeds and exotic plants: relationships to disturbance in mixed-grass prairie
✓ Methylamine/ammonium uptake systems in *Saccharomyces cerevisiae*: multiplicity and regulation
✓ Resistance to infection with intra-cellular parasites – identification of a candidate gene
✓ Mass flux measurements at active lava lakes: Implications for magma recycling

10.3 Strategy 3: Choose strategically – noun phrase, statement, or question?

The traditional way to write titles and headings is as a noun phrase: a number of words clustered around one important "head" noun. Examples of this kind of title include the following, where the head nouns are shown in bold:

- **Diversity and invasibility** of southern Appalachian plant communities
- Food expenditure **patterns** in urban and rural Indonesia
- **Systems** of weed control in peanuts
- Iron **uptake** by symbiosomes from soybean root nodules
- **Evidence** of involvement of proteinaceous toxins from *Pyrenophora teres* in net blotch of barley

Several of these titles are very effective: brief, informative, and with keywords placed near the front. However, this style of title writing is not always the best for meeting the two guidelines discussed under Strategies 1 and 2. Look again at the last title in the list, "Evidence of involvement of proteinaceous toxins from *Pyrenophora teres* in net blotch of barley." This title leaves us with an unanswered question: what kind of involvement? Additionally, the first four words are very general in meaning, giving no enticement to the reader to continue reading. Rewriting this title as a statement could overcome these difficulties, and was in fact recommended by a referee when this paper was under review. (A statement is a sentence with a subject and a verb, and its advantage in this context is that it can give more explicit information about the results of the study.)

✗ Evidence of involvement of proteinaceous toxins from *Pyrenophora teres* in net blotch of barley
✓ Proteinaceous metabolites from *Pyrenophora teres* contribute to symptom development of barley net blotch (Sarpeleh et al. 2007)

Statement titles are only suitable for papers that address a specific question and present a non-complex answer. In these conditions, the sentence form is a good option to replace titles that begin with vague terms such as "The effects of. . ." For example:

✗ Effects of added calcium on salinity tolerance of tomato
✓ Calcium addition improves salinity tolerance of tomato

When there is no simple answer to be presented, it can be effective to write a title as a question. For example:

✓ Which insect introductions succeed and which fail?

As with all sections of your manuscript, check whether your target journal has specific conventions or recommendations about the form and content of titles before you decide which form to use. Research across recent issues can be helpful for investigating titling preferences, including word count and which elements of content to include, as well as what form to write in; investigating 50–75 titles can give an accurate indication (Kerans et al. 2016, 2020).

In our own experience, it can be useful to develop a list of possible titles as you draft your manuscripts, and choose the most effective one for the target audience and the paper's key message right at the end of the writing process.

10.4 Strategy 4: Avoid ambiguity in noun phrases

If writers place a string of nouns and adjectives together to form a title which packs a lot of meaning into a few words, they can sometimes cause problems of ambiguity: more than one possible meaning. This is particularly the case when nouns are used as adjectives; that is, when they are placed in front of the head word of the noun phrase. To investigate why this is so, let us consider some examples.

The noun phrase "germination conditions" has only one possible meaning – conditions for germination – and thus it can be used without risk of ambiguity. Similarly, "application rate" can only mean the rate of application. However, "enzymatic activity suppression" could mean either suppression *of* enzymatic activity or suppression *by* enzymatic activity and is therefore ambiguous. A general guideline is to restrict these noun phrases to a maximum of three words, and then only if there is no risk of misunderstanding. If such phrases grow longer, rewrite them by inserting the prepositions that clarify the meaning (e.g. *of, by, for*). For example:

✗ soybean seedling growth suppression
✓ suppression of soybean seedling growth

N.B. When nouns are used as adjectives in extended noun phrases, they are always used in the singular. Useful examples to help you remember this are as follows:

food for dogs	→	dog food
disturbance by herbivores	→	herbivore disturbance
nodules on soybean roots	→	soybean root nodules

To practice application of the four strategies, complete Task 10.1, which is an exercise in analysing the structure and communicative effectiveness of selected article titles.

Task 10.1   Analysing article titles

Complete Table 10.1 for each of the PEAs and your Selected Article (SA) and discuss your findings with a colleague or teacher, if appropriate. Compare your answers with our suggestions in the Answer pages.

Table 10.1 Analysing article titles.

Question	Kaiser et al. (2003)	Britton-Simmons & Abbott (2008)	Ganci et al. (2012)	Your SA
Is the title a noun phrase, a sentence, or a question? How many words are used in the title? What is the first idea in the title? Why do you think this idea has been placed first?				

Now, spend a little time deciding if there are any improvements you can make to the title you have drafted for your Own Article (OA), if appropriate.

CHAPTER 11

The Abstract and highlights

11.1 Why Abstracts are so important

- For busy readers, the Abstract, sometimes called the Summary, may be the only part of the paper read, unless it succeeds in convincing them to download and take the time to read the whole paper.
- For readers with limited access to the literature (working outside institutions with comprehensive journal subscriptions), the Abstract may be the only information on your work that is available.
- Abstracting services may use the text of the title plus the Abstract and keywords for their searchable databases.

11.2 Selecting additional keywords

You will be familiar with the term "keyword" because of its relevance to constructing compelling titles (see Chapter 10). Consult other similar papers in your field to see which additional keywords they use beyond the ones already included in the title. The idea is to select words or short phrases that represent key concepts or features of your study. At this stage, think again about your audience and their interests and try to predict what keywords they might use to search under. Do a keyword search using the keywords you are considering, and see if it retrieves papers similar to your own. Check also for guidance provided in the Instructions to Contributors of your target journal.

An additional important use of an article's keywords is by the editor of the journal it is submitted to. The editor is likely to use the keywords to match with those provided by prospective referees to describe their particular expertise. Therefore, it is a good idea to ensure that the keywords you list describe the expertise you would like the reviewers of your paper to have.

11.3 Abstracts: typical information elements

Some journals provide a list of questions or headings for authors to respond to in writing their Abstracts (structured abstracts), and others do not. All provide a maximum number of words that an Abstract (or Summary) may contain (e.g. 250 for

Writing Scientific Research Articles: Strategy and Steps, Third Edition.
Margaret Cargill and Patrick O'Connor.
© 2021 John Wiley & Sons Ltd. Published 2021 by John Wiley & Sons Ltd.

The Plant Journal and 350 for the *Journal of Ecology*, as of September 2020). Based on analyses of many Abstracts in science and technology fields, the following information elements can be proposed as constituting a full abstract or summary (Weissberg & Buker 1990):

Some background information	B
The principal activity (or purpose) of the study and its scope	P
Some information about the methods used in the study	M
The most important results of the study	R
A statement of conclusion or recommendation	C

This list is often compressed to the following components:

Principal activity/purpose and method of the study	P + M
Results	R
Conclusion (and recommendations)	C

Complete Task 11.1.

Task 11.1   Analysing Abstracts/Summaries

Read the Abstracts/Summaries of all three of the Provided Example Articles (PEAs) and identify which of the information elements listed are present, and in which sentence(s). (Even if you are not completely familiar with the science being presented in the papers, these sections are short enough that you should be able to complete this task without difficulty, and there are important things to learn from doing so.)

Compare your answers with our suggestions in the Answer pages.

N.B. The *Journal of Ecology*, which published Britton-Simmons & Abbott (2008), provides the following guidelines for the writing of the Abstract:

The Abstract should be written to be understood in isolation from the rest of the paper and must not exceed 350 words. Using typically four simple, factual, numbered statements, it should:

- give the conceptual context
- state the methodological approach
- report the main results and conclusions
- include a final point headed "Synthesis", which sums up the paper's key message in generic terms that can be understood by non-specialists, indicating clearly how this study has advanced ecological understanding. (https://besjournals.onlinelibrary.wiley.com/hub/journal/13652745/author-guidelines)

The final sentence of this advice is particularly relevant to us in our analysis of this paper, as it provides a rationale for what has been emphasised in the strategically important parts of the paper: the Title, the Summary (now called Abstract in this journal), the end of the Introduction, and the Discussion. This fact underlines how very important it is to seek out, read carefully, and respond effectively to the Author Guidelines (or equivalent) for the journal to which you will submit your manuscript.

Now complete Tasks 11.2 and 11.3, if appropriate.

> **Task 11.2   Analysing your SA Abstract or Summary**
>
> Repeat Task 11.1 for your Selected Article (SA) and discuss your findings with a colleague or teacher, if appropriate.

> **Task 11.3   Drafting your own Abstract or Summary**
>
> Now write or revise your own Abstract or Summary, if appropriate. One way to begin is to write sentences for all of the information elements just listed and combine them into a first draft. Then check the number of words you have used against the requirement of the journal you are targeting. If necessary, shorten your draft, using techniques such as those you have observed in the Abstracts/ Summaries you have analysed.

11.4 Visual abstracts

Visual abstracts are becoming very common as a way of graphically summarising the information from a written Abstract. They have particular currency in article promotions on social media and provide an additional method for conveying key messages and results in a brief format. The guidance provided in Chapter 5 for best practice visualisation is also relevant for visual abstracts, but you should consult the Instructions to Contributors of the journal you are targeting and consider the following points.

To convey a clear message, it is necessary to focus on the experience of the reader at all stages of development. This requires that you have a clear understanding of what message you are seeking to convey, even if the message needs to be simplified for visual presentation. The visual abstract is a companion to other components of the research article, not a substitute, and readers can refer to the detail even if they first encounter a simplified version in the visual abstract. When you compromise for simplicity, compromise completeness, not the main message (i.e. it is better to sacrifice some detail than to lose the larger message being conveyed).

One method of improving your design is to recognise that many changes to the first draft may be needed and that every step towards partial improvement is a good step – you are unlikely to jump to a perfect product in one go. Share your drafts with others and take feedback on board. Be prepared to be creative, but not for the sake of it; you have put a lot of work into your research article and you want the visual abstract to get the attention of the right audience for the right reason.

11.5 "Highlights" and other significance or summary sections

Highlights are bullet points that aim to convey the core findings of your article and enhance its online discoverability via search engines. They should aim to capture the novel results and/or new methods in the research. Try to include terms that your anticipated readers would be searching for, and do not try to capture all the ideas of the manuscript. Highlights will usually be three to five bullet points (depending on the journal), with each being no more than 85 characters in length (refer to the Instructions to Contributors for your target journal; note that some may name this kind of section differently).

CHAPTER 12

Writing review articles

Although the main focus of this book is the writing of research articles, we include here a discussion of writing review articles, for the sake of completeness and to provide a basis for comparing the two article types.

Review articles can make a strong contribution to a scientist's reputation and citation metrics. They are often highly cited, and if written early in a career can represent the "balance-tipper" that leads to getting an interview for a desired job. Some important journals are devoted to reviews or have a dedicated review section. Reviews in many journals are commissioned or solicited (i.e. authors are invited to prepare and submit a review on a particular topic), but these are still subject to the normal process of peer review. Another potential route is for authors to contact the editor of an appropriate journal and ask if the journal would be interested in a review on the topic they wish to write on. Whatever the way in, it is very important that the submitted review is comprehensive and timely, presents a compelling argument, and is well structured and written. This chapter aims to provide guidance for achieving these goals. The chapter also addresses some of the important structural features of a *systematic review*, a type of review which is formally structured, may or may not accompany a meta-analysis, and develops and presents the methodology of the review along with the results. Systematic reviews are becoming more common in a range of fields of research, in order to assist in the synthesis of current knowledge around a defined question (or questions). They can be highly desirable as a starting point for identification of knowledge gaps and for policy or practice development.

Most scientist authors gain their first experience with this genre, or type of article, through writing a "literature review," first to support their PhD or Masters project, and then as a chapter of their thesis. Some thesis literature reviews have the potential to be reworked after the fact as review articles for the international literature, or they may be written with submission in mind from the beginning. Such a broadening of audience often requires both a shift in thinking about the aims of presenting the material (Box 12.1) and a more sophisticated application of the scientific writing principles presented elsewhere in this book (see Chapter 8). Of course, adopting these "shifts" during the thesis-writing process itself can only be beneficial for the quality of the thesis, and potentially the outcome of the examination. An interview study with experienced thesis examiners (Mullins & Kiley 2002) identified that initial impressions of thesis quality are often formed by the end of the literature review. Thus, it is important for scientists to develop skills for writing (and supervising) effective, targeted reviews of literature in a particular subfield.

Writing Scientific Research Articles: Strategy and Steps, Third Edition.
Margaret Cargill and Patrick O'Connor.
© 2021 John Wiley & Sons Ltd. Published 2021 by John Wiley & Sons Ltd.

Box 12.1 Aims of reviews: differences between project-focused and journal-targeted reviews

In order to highlight key points of difference for reviews destined for publication, it is instructive to revisit what higher education literature and practice have to say about the aims of literature reviews focused on specific research projects, as for a PhD project or indeed a grant proposal. To put this examination in context, we should also focus on the different audiences and assessment criteria for the two types of reviews (Table 12.1).

Table 12.1 Differences between project-focused and journal-targeted reviews.

	Project-focused reviews	Review articles
Audience	Academic supervisors or committees; thesis examiners; funding agencies	The journal's editor, and subsequently its referees, reading with the perspective of the journal's readers
Assessment criteria	Academic rigour and persuasiveness: Does the review provide strongly argued and substantiated justification for conducting the research project that is proposed?	Novelty and significance of the contribution; this may include proposed directions for future research, but will not generally include details of a specific project to be undertaken
Relevant aims	"The chain (1)" (Weissberg & Buker 1990) "To provide background information needed to understand your study; To show familiarity with the important research which has been done in your area; To establish your research as one link in a chain which is developing and expanding knowledge in your area." "The chain (2)" (Pechenik 1993). In the review of literature, "you review the primary literature on a particular topic, but you do so with a particular goal in mind: you wish to lead your reader to the inescapable conclusion that the question you propose to address follows logically from the research that has gone before." "The gap(s)" (2) A literature review should serve to support, explain, and illuminate the logic behind the proposed research; it is used to explain the choices you have made in your research.	"The argument" (Lindsay 1995, 2011) "The purpose of a good review is not to present a catalogue of names, dates and facts, but to present reasoned arguments about the field under review based on as many names, dates, and facts as are necessary to support these arguments." "The gap(s)" (1). The review article is a critical analysis of relevant sources which shows what knowledge is still to be understood, or what research still needs to be done.

This chapter draws on the perspectives of authors (O'Connor and interviewees), referees, and publishers of review articles, and on experience gained in teaching workshops ("Literature reviews and review papers: constructing compelling arguments") to research students engaged in writing or redrafting review papers or chapters (Cargill). Rather than including a single example review in the book, we have chosen to focus on the extensive collection of well-regarded Tansley review articles, made freely available on the website of the New Phytologist Trust (www.newphytologist.org/reviews). Readers can begin with our analysis of key features, identify them in the short extracts we quote here, and then download whichever of the examples seem most relevant for more in-depth investigation. Alternatively, a recent review from a relevant field or a potential target journal can be used to conduct the analyses we suggest.

12.1 What editors want to publish

When asked for advice on writing reviews suitable for the *New Phytologist*, its managing editor gave this as her initial comment: that unsuccessful authors often "do not provide any new synthesis or conclusions that change our under-standing since the last literature review" (H. Slater, pers. comm., June 2012). This should be considered alongside the guidelines given for the (shorter) research review category in the same journal: "Following a short introduction putting the area into context, and providing a 'way in' for the nonspecialist, these will concentrate on the most recent developments in the field" (www.newphytologist.com/authors). Taken together, these guidelines emphasise three features: an audience-responsive introduction, up-to-date coverage of the relevant work, and synthesis and conclusions that add to the previous understanding of the field; that is, "new ideas" (Lindsay 1995, 2011). We begin with the last of these points, and focus first on applying the concept of the "take-home message" (THM) to the review genre.

12.2 The "take-home message" of a review

We have used this concept extensively in the previous chapters dealing with research articles, including as an organisational tool to assist authors in selecting and presenting their research data so that they all support a common message or conclusion. A similar process is applicable for review articles, although the "data" are different. It can be useful to think of these "data" as your critical evaluations of the previous work you are reviewing; thus, you need to arrange and present them so that they support a clear and consistent message. Potential processes for developing the THM are worth some thought.

The development of THMs when planning and writing a review can occur in several ways, often in conjunction with planning the order of the sections:

- If you have a firm idea of the argument to be developed and the desired THM, as was the case with a review by Harris et al. (2011) (John Harris, pers. comm., July 2012), then it is effective to create a set of bullets that capture the main points and use these as the basis of the first draft of the Table of Contents. The task is then to collect and arrange the evidence to support each point and link it together logically.

- If the evidence is largely collected but the THM is still elusive, it is useful to sort the evidence under topic headings. Lindsay (2011) suggests that, for each one, authors should note both the topic to be dealt with and the conclusion they want to make about it.
- If the proposed THM is clear but the structure of the argument is not, then a bottom-up process can be useful. Write the THM at the bottom of a sheet of paper, and then work your way upwards, answering this question for each entry: What does my reader need to know in order to believe that conclusion? Some work will be needed to produce a logically ordered set of headings – but once it is done, the process of writing the paragraphs should be more straightforward.

It can then be useful to think about where in published reviews the THM appears. A subsidiary question is whether it appears all at once, or whether aspects of it can appear in different places, to scaffold the development of the argument and then reinforce it at the end. The potential places are these: title, Abstract, Table of Contents, the end of the Introduction, and Conclusions. See Table 12.2 for a comparison of four of these text sections from two Tansley reviews: Harris et al. (2011) and Wymore et al. (2011). The full outlines of the articles are given in Boxes 12.2 and 12.3.

Table 12.2 Locations of elements of THM in two recent reviews. Highlighting identifies text deemed particularly relevant to the THMs.

Location	Harris et al. (2011)	Wymore et al. (2011)
Title	Modulation of plant growth by HD-Zip class I and II transcription factors in response to environmental stimuli	Genes to ecosystems: exploring the frontiers of ecology with one of the smallest biological units
Abstract	Plant development is adapted to changing environmental conditions for optimizing growth. This developmental adaptation is influenced by signals from the environment, which act as stimuli and may include submergence and fluctuations in water status, light conditions, nutrient status, temperature and the concentrations of toxic compounds. The homeodomainleucine zipper (HD-Zip) I and HD-Zip II transcription factor networks regulate these plant growth adaptation responses through integration of developmental and environmental cues. Evidence is emerging that these transcription factors are integrated with phytohormone-regulated developmental networks, enabling	Genes and their expression levels in individual species can structure whole communities and affect ecosystem processes. Although much has been written about community and ecosystem phenotypes with a few model systems, such as poplar and goldenrod, here we explore the potential application of a community genetics approach with systems involving invasive species, climate change and pollution. We argue that community genetics can reveal patterns and processes that otherwise might remain undetected. To further facilitate the community genetics or

Table 12.2 (*Continued*)

81

Writing review articles

Location	Harris et al. (2011)	Wymore et al. (2011)
	environmental stimuli to influence the generically preprogrammed developmental progression. Dependent on the prevailing conditions, adaptation of mature and nascent organs is controlled by HD-Zip I and HD-Zip II transcription factors through suppression or promotion of cell multiplication, differentiation and expansion to regulate targeted growth. In vitro; assays have shown that, within family I or family II, homo-and /or heterodimerization between leucine zipper domains is a prerequisite for DNA binding. Further, both families bind similar 9-bp pseudopalindromic cis elements, CAATNATTG, under in vitro; conditions. However, the mechanisms that regulate the transcriptional activity of HD-Zip I and HD-Zip II transcription factors in vivo; are largely unknown. The in planta implications of these protein–protein associations and the similarities in cis element binding are not clear.	genes-to-ecosystem concept, we propose four community genetics postulates that allow for the conclusion of a causal relationship between the gene and its effect on the ecosystem. Although most current studies do not satisfy these criteria completely, several come close and, in so doing, begin to provide a genetic-based understanding of communities and ecosystems, as well as a sound basis for conservation and management practices.
End of the Introduction	In this review, we present our account of the current knowledge that relates to the roles of the HD-Zip I and HD-Zip II TFs during plant adaptation under changing environmental conditions. We also briefly describe how these TFs integrate with phytohormonemediated responses, and indicate the limits of the current knowledge that relate to the mechanism of transcriptional activity and address the issue of how to overcome these limitations.	The major goal of this review is to explore how this concept applies to systems for which this approach has not been explicitly employed, yet, are sufficiently developed to explore broader basic and applied issues. We develop our ideas in the context of global change associated with commonly occurring, ecosystemimpacting events, including invasive species, climate and pollution. For example, in conifers, we explore how the interactions of foundation species (trees and squirrels) and climate can affect a much larger community. With examples from two highly invasive species that have become foundation species in their new environments, we explore how a single mutation in one example and a single

(Continued)

Table 12.2 (*Continued*)

Location	Harris et al. (2011)	Wymore et al. (2011)
		haplotype in another example can have cascading effects to redefine their respective ecosystems. Similarly, with the release of endocrine-disrupting chemicals from human contraceptives into aquatic ecosystems, we explore how pollution can alter the gene expression of foundation species, which, in turn, may redefine these ecosystems. Thus, a community genetics perspective on interacting foundation species, exotics and pollution can broaden our understanding of how the genetics of foundation species can have unexpected consequences, and remind us of the complex connections that exist in both natural and exotic systems.
Conclusions	Plants cope with a variety of environmental stresses by modifying their growth pattern to minimize the impacts of stress or to escape damage. The HD-Zip I and HD-Zip II TFs play an integral role in the signaling network that is triggered by endogenous and external stimuli, which leads to the modified growth characteristic of stressed plants (Fig. 3). This growth adaptation is achieved through the regulation of cell differentiation, division and expansion by HD-Zip I and HD-Zip II TFs. However, there are still mechanistic and functional aspects of the HD-Zip I and HD-Zip II network that we need to understand before we are able to characterize the roles of each family in plant growth adaptation to environmental stresses. We know very little of the downstream genes that are ultimately regulated by the HD-Zip I and HD-Zip II TFs. Identification of their target genes is needed to build a comprehensive view of the regulatory pathways and will enable validation of	In summary, it seems that John Muir (1911) might have been correct when he stated that, 'When we try to pick out anything by itself, we find it hitched to everything else in the Universe.' Numerous examples have emerged from diverse systems that make the case that the genes-to-ecosystem approach can provide an important perspective for the understanding of complex systems, for informing land managers and even for evaluating the effect on the human condition where the genetic impacts of pollution can have unintended effects on the food supply and human health. Similar to Koch's postulates, we present four community genetics postulates for confirming or rejecting the hypothesis of a genetic effect on the community and ecosystem (Table 1): (1) the demonstration of a target species' impact on the community and ecosystem; (2) the demonstration

Table 12.2 (*Continued*)

Location	Harris et al. (2011)	Wymore et al. (2011)
	the suite of promoters controlled by HD-Zip I and HD-Zip II proteins to define the cis-acting elements. More information is needed on the interaction profiles of HD-Zip I and HD-Zip II TFs, and the properties that specify interaction partners which contribute to cis element binding. The increasing amount of cell-specific microarray data available in public databases will also enable determination of potential interaction partners that are dependent upon expression in the same cell. If transcriptional regulation by the HD-Zip I and HD-Zip II TFs is conferred through a common cis element, that implies that there are factors determining the specificity of transcriptional activity, as transgenic plants over- or under-expressing HD-Zip I and HD-Zip II TFs have variant phenotypes. Elements contributing to the specificity of action remain elusive. Obtaining data on the three-dimensional structures of the HD-Zip I and HD-Zip II TFs will also decisively contribute to our understanding of the nature of the protein–protein and protein–DNA interactions.	of key traits that are heritable; (3) the demonstration of genotypic variation in the communities they support and ecosystem processes; and (4) the manipulation of target gene(s) or their expression to experimentally evaluate a community and ecosystem effect. The last of these is the least well documented (but see the exotic hydrilla example of Michel et al., 2004; Fig. 2). Nevertheless, as we genetically engineer and release organisms, the fourth postulate will be evaluated on a global scale. In complex systems involving many interacting species, we believe that there are three main advantages to this approach. First, the incorporation of a genetically based model places community and ecosystem ecology within an evolutionary framework subject to natural selection. Second, because a genes-to-ecosystem approach studies species within a community context, it is more realistic and less likely to result in management errors compared with a single species' approach. Third, the use of the genes-to-ecosystem concept can reveal important interspecific indirect genetic effects among species, thus generating meaningful applications for the conservation of biodiversity, restoration, bioengineering, climate change and even the understanding of important human diseases.

Box 12.2 Outline of "Modulation of plant growth by HD-Zip class I and II transcription factors in response to environmental stimuli" (Harris et al. 2011)

Summary

I Introduction
  1 Plant development is adapted to the environment
  2 Discovery of the homeodomain transcription factors
  3 Structure of the homeodomain helix-turn-helix, the role of the leucine zipper in HD-Zip proteins and three-dimensional structures of nonplant HD and leucine zippers

II The role of HD-Zip transcription factors in plant growth adaptation to environmental changes and the phytohormone network
  1 The role of HD-Zip transcription factors in water deficit, salinity stress and ABA-modulated development
    HD-Zip I
    HD-Zip II
  2 The role of HD-Zip transcription factors in plant growth adaptation to light and their expression and function
    HD-Zip I
    HD-Zip II
  3 Integration of endogenous and environmental signaling through phytohormones and the HD-Zip I and HD-Zip II transcription factors

III Dissecting the common cis element, dimerization and cell specificity of HD-Zip I and HD-Zip II transcription factors
  1 Defining target cis elements within downstream genes
  2 Regulation of cis element binding through posttranslational modification
  3 Dimerization and the roles of members of the HD-Zip I and HD-Zip II families in the cell- and condition-specific interactome
  4 Redundancies in the roles of HD-Zip I and HD-Zip II transcription factor paralogs

IV Conclusions

Acknowledgements

References

Source: Based on Harris, J. C., Hrmova, M., Lopato, S., & Langridge, P. (2011). Modulation of plant growth by HD-Zip class I and II transcription factors in response to environmental stimuli. *New Phytologist*, 190(4), 823–837.

Box 12.3 Outline of "Genes to ecosystems: exploring the frontiers of ecology with one of the smallest biological units" (Wymore et al. 2011)

Summary
I   Introduction
II   Fundamental principles and the community genetics equivalent of Koch's postulates
III Genes, invasions and competition
IV   Mutation, resistance and ecosystem consequences
V   Heritable traits, pine cones and climate
VI   Gene expression, fish and pollution
VII An emphasis on foundation species and their biotic and abiotic interactions

VIII Applications to the human condition

IX Conclusions

Acknowledgements

References

Source: Wymore, A. S., Keeley, A. T. H., Yturralde, K. M., Schroer, M. L., Propper, C. R., & Whitham, T. G. (2011). Genes to ecosystems: exploring the frontiers of ecology with one of the smallest biological units. *New Phytologist*, 191(1), 19–36.

Commentary on Table 12.2: effective highlighting of THMs

Titles

In both Harris et al. (2011) and Wymore et al. (2011), the title is a noun phrase (i.e. not a sentence) and thus names a topic rather than stating a conclusion. This seems to be by far the most common format for review titles, although questions can also be included (examples from the recent Tansley list are "Units of nature or processes across scales? The ecosystem concept at age 75" and "Innate immunity: has poplar made its BED?"). The Wymore title gives clues to its claim for significance in its vocabulary (*frontiers of ecology*) and grammatical form (*exploring* – a verbal noun (gerund), which gives more sense of action than the corresponding noun form *exploration of*). The Harris title focuses on the three parameters of its coverage: *modulation of plant growth*; *by* (particular actors); and *in response to* (particular stimuli). If titles are the advertising banner for an article, as suggested in Chapter 10, then these two articles have taken different routes to attract their readers: appetite whetting and accurate content delineation. Choosing the most effective route for your own review title can be a lot of fun – we suggest writing a list of potential titles as the writing proceeds and making the ultimate selection once the article is in its final form.

Abstract

The stages used in Chapter 11 for analysing the Abstracts of research articles may need revision for application to review Abstracts. To recap, these stages are Background, Purpose and scope, Method(ology), Results, and Conclusion/recommendation. Complete Task 12.1 now.

Task 12.1 Analysis of review Abstracts

Read the Abstracts of the two reviews presented in Table 12.2. For each, identify any sentences that seem to you to match any of the five stages from Chapter 11: Background, Purpose and scope, Method(ology), Results, and Conclusion/recommendation.

Check your answers with our suggestions in the Answer pages.

You will see from our suggested answers to Task 12.1 that *Background* appears in both examples and *Purpose and scope* is set out in one of the two. *Method(ology)* is mentioned in neither of the two examples, but that will not be the case for reviews in all discipline areas – for example, systematic reviews in medicine often require a focus on this element. We have identified sentences that reflect *Results* in both examples and have described this element as "information representing new synthesis or conclusions." *Conclusions/recommendations* also occurs in both examples, in one case highlighting remaining gaps in knowledge, in the other making a claim for the significance of the new synthesis/conclusions presented in the paper. We suggest that a usable analysis framework for review article Abstracts consists of the following "stages":

- Background (information available and accepted in the field)
- Purpose and scope (optional – may include points of differentiation from previous reviews)
- Method(ology) (if required by review type)
- Summary of new synthesis or conclusions reached in the review
- Conclusion (remaining knowledge gaps, significance of new synthesis, etc.)

This framework will be useful for analysing review Abstracts in your own field, to check for discipline-specific variation in how THMs are highlighted, and also as a checklist for writing the Abstract of your own review article.

Discussion of the Table of Contents can be found in Section 12.3.

End of the Introduction

In Chapter 8, we emphasise the importance of this segment of a paper and characterise it as Stage 4 in the argument structure of an Introduction, with three possible forms: a statement of aim or purpose of the present work; a statement of the main activity of the paper (what the paper does); or a summary of the findings of the paper – depending on the subfield and journal conventions. We also suggest that this stage is written early in the article drafting process, and as the first step in writing the article Introduction. All these suggestions apply equally to the writing of review papers. Now look at Table 12.2 again and complete Task 12.2, analysing Stage 4 of each article.

Task 12.2   Analysis of ends of Introductions in reviews

For the two end of the Introduction sections in Table 12.2, identify:

1 which of the following options for Stage 4 are present (more than one option may be present in each case):
   a. statement of aim or purpose of the present work;
   b. statement of main activity of the paper (what the paper does, or what the authors do in the paper);
   c. summary of the findings or outcomes of the paper;
2 which words or phrases show that a particular option is being used; and
3 whether the section also functions as a map of how the rest of the article is arranged (Introduction Stage 6, Figure 8.1), and if so, which words signal this function.

Check your answers with our suggestions in the Answer pages.

Table 12.2 uses grey highlighting to identify parts of the text that relate particularly to the THM of the review. Of interest, the words and phrases identified in Task 12.2, Part 2 are not highlighted in Table 12.2 – it is the noun phrases (e.g. *the current knowledge that relates to the roles of the HD-Zip I and HD-Zip II TFs during plant adaptation under changing environmental conditions*) and noun clauses (e.g. *how this concept applies to systems for which this approach has not been explicitly employed, yet, are sufficiently developed to explore broader basic and applied issues*) that come after these words which carry the main elements of the THM. This analysis demonstrates how a Stage 6 "map" can be written so that it is not mundane or uninteresting, but instead contributes to the delineation of the article's THM.

Conclusions

As indicated by the diagram shape used to represent AIBC (Abstract, Introduction, Body sections, Conclusions) articles in Chapter 2 (Figure 2.4), the final Conclusions section is expected to relate clearly to the article's Introduction. Table 12.2 provides an opportunity to compare these sections at a glance for two articles. For each article, or for the review from your own field you are using for analysis, consider the following questions:

- Are all the issues raised or points made in the end of the Introduction reflected in some way in the Conclusions?
- Are there issues or points made in the Conclusions that are not foreshadowed in the end of the Introduction?
- If there are any mismatches, do they have a negative effect on the article's logical flow – or on the strength of the argument produced?

Borrowing from the checklist in Chapter 9 for the content of Discussion sections of research articles, we can suggest that the Conclusions of a review could also be expected to contain information relevant to these three elements, if applicable:

- Limitations of the study
- Implications of the study (here, of the new synthesis and conclusions)
- Recommendations for future research and/or practical applications

Harris et al. (2011) clearly focus on the second (implications: *If . . . then . . .*) and third (future research directions) of these points. Wymore et al. (2011) stress implications, including in the guise of advantages, and also address limitations (*The last of these is the least well documented. Nevertheless. . .*).

A final issue that sometimes causes concern for novice authors is ensuring that the wording is not overly repetitive between the Abstract and the Conclusions. A close comparison, such as that presented in Table 12.2, can provide examples of effective rewordings that meet the need for originality while making it clear that the same concepts are under discussion. English as an additional language (EAL) authors may wish to conduct similar comparisons using reviews from well-respected journals in their own fields, to further build the range of appropriate wordings they have at their disposal.

12.3 The structure of review articles

Review articles follow the AIBC structure depicted in Figure 2.4. To summarise, reviews in the sciences generally have an Introduction and a Conclusions section, with similar functions to those found in research articles (but see the commentary on Table 12.2 for important specific features). The central Body section is divided into themed subsections, each with a heading (and sometimes a number of subheadings) that indicates to the reader the content to be covered (see Boxes 12.2 and 12.3 for examples). If these headings and subheadings are written with care, the main points of the argument can be effectively highlighted as well.

In this regard, it is worthwhile noting how many levels of heading are included in the Contents section that is commonly presented before the Summary or Abstract in the journal publishing the review. In the *New Phytologist* only the top level is included. This means that, in the Harris et al. (2011) review (Box 12.2), the richness included in the second level of heading is not brought to the readers' attention at first glance. In contrast, the Wymore et al. (2011) review (Box 12.3) relies on only one level of heading, and the full story is therefore available to the readers "up front." Even if the journal does not publish the Contents separately, experienced readers often skim an article, reading only the headings, to get an overall view of the content and argument, so it is worthwhile to ensure that the headings present an effective "roadmap" or "scaffolding" for the article.

Topic sentences in the Body sections: a key to signalling logical flow

As per Strategy 1 in Section 8.9, topic sentences at the start of Body paragraphs are an excellent way to introduce a point or foreshadow the direction of an argument. In the context of a review article, they can add considerably to the effectiveness of the "scaffolding" provided by the section headings. To demonstrate how this can work, we present in Box 12.4 the first sentences only of the paragraphs of a single Body section of the Wymore et al. (2011) review. For those who have already skimmed the key THM locations (Table 12.2), as readers of this chapter will have done, reading just the contents of Box 12.4 enables a good grasp of the argument presented by this section of the review.

Box 12.4 Heading and topic sentences from a Body section of Wymore et al. (2011)

V Heritable traits, pine cones and climate
  The level of serotiny, a heritable trait in lodgepole pine (*Pinus contorta*) stands, is influenced by climate, fire and seed predators, and, in turn, affects forest composition and dependent species' evolution (Fig. 3). . . .
  Climate, through its effect on fire regimes, appears to exert a major selection pressure that acts on serotiny. . . .
    Seed predators also influence the level of serotiny in lodgepole pine stands (Fig. 3b). . . .
  The interaction of fire, herbivory and serotiny cascades to affect the whole forest ecosystem through sapling density after a fire. . . .
  Geographic location and pre-fire levels of serotiny also explained much of the observed variation in biotic responses, including species' richness, abundance of opportunistic species, and cover and density of pine seedlings, forbs, graminoids and shrubs post-fire (Fig. 3d; Turner et al. 1997). . . .
    The level of serotiny in a population also affects the evolution of individual species. . . .
    To summarize, climate affects fire regimes, which, together with seed predators, select for or against the heritable trait of serotiny in lodgepole pine stands. . . .
  The lodgepole pine system fulfills three of the four postulates.

Source: Wymore, A. S., Keeley, A. T. H., Yturralde, K. M., Schroer, M. L., Propper, C. R., & Whitham, T. G. (2011). Genes to ecosystems: exploring the frontiers of ecology with one of the smallest biological units. New Phytologist, 191(1), 19–36.

Effective topic sentences do not always flow from authors' fingertips as they write the first draft of a review – they can be added and refined during the revision process, often in response to the question, "What point am I trying to make here?"

12.4 Visual elements in review articles: tables, figures, and boxes

Visual elements can be very effective in clarifying and summarising the outcomes of new analysis in a review, or presenting compelling overviews of newly proposed mechanisms or processes. As is emphasised in Chapter 5, having a clear THM for each visual element is very important, and the table title or figure legend should be carefully crafted to highlight it. Our selected Tansley reviews provide examples of the first two categories, and additional reviews are also cited that demonstrate specific features.

Tables in reviews

Careful design of a table and crafting of its column headings can enable very effective summarising of large amounts of data to make a particular point. For example, in Kranner et al. (2010), a table has been included with the following features:

Title: Table 1 Examples of potential abiotic stress factors and their effects on whole plants and orthodox seeds, classified according to the eustress–distress concept
Column headings: Stress factor; Effect on whole plants (Distress/Eustress); Effect on orthodox seeds (Distress/Eustress)
Row labels: Water deficit; Temperature; Fire; Nutrients; Wind; Contamination, for example by non-essential heavy metals

The first reference (of four) to the table in the text is as follows:

As a result of their essential role in plant reproduction, one would intuitively expect that plants have evolved mechanisms that protect their seeds from stress.

Indeed, in the dry, quiescent state, protected by their seed coat, many seeds are exceptionally tolerant of stress factors, such as temperature extremes, that are lethal to adult plants (Table 1).

The second reference reads like this:

In addition, severe stresses on the mother plant will generally cause distress for both orthodox and recalcitrant seeds (Table 1; Fig. 3).

Thus, evidence can be referred to in the developing argument as required without repeating details, and the citations that support each piece of evidence are included in the table cells.

Another effective use of a table is demonstrated in the Wymore et al. (2011) review. Its Table 1 presents the four postulates the authors have developed to establish a causal relationship between genes and their community and ecosystem consequences. Thus, whenever they wish to refer to one or more postulate, they reference the table so that all the appropriate detail is easily available to readers.

Figures in reviews

Two main types of figures are used in review articles:

- composite figures presenting several instances of data (with citations) as evidence for a point being made in the review (e.g. Figures 3 and 4 in Wymore et al. 2011); and
- flow charts or schematic diagrams showing steps in a newly proposed process or mechanism (e.g. Figure 3 in Harris et al. 2011, p. 830).

In both cases, the figure legends need to be comprehensive, ensuring that the figure can be understood without reading the associated text, and appropriate citations must be included. The figure legend of the Harris et al. (2011) example is as follows:

Fig. 3 The proposed plant developmental regulatory network involving homeodomainleucine zipper (HD-Zip) I and HD-Zip II transcription factors (TFs). Black arrows indicate an interaction. The black dashed line represents a promoter, the cross-hatched box represents cis elements downstream of other trans-acting factors, the CAATNATTG box represents the HD-Zip I and HD-Zip II cis elements, the right-angled arrow represents the transcriptional start site, and the 'Downstream gene' box represents genes whose transcription will be either suppressed or activated. The arrowed box represents post-translational modification that affects the ability of the HD-Zip TFs to dimerize and/or bind DNA (Himmelbach et al., 2002; Tron et al., 2002).

The text reference to the figure is as follows:

These observations suggest that a large network exists, where dimerization partners confer different transcriptional characters and two families compete over similar cis elements. This would enable environmental and endogenous signals to regulate fluxes in the network to establish developmental programs through differential gene expression (Fig. 3).

The verb forms in the text (*suggest*, *would enable*) combine with the word choices in the legend (*the proposed plant developmental regulatory network*. . .) to clearly identify the status of the network, and its depiction in a figure enables it to be inserted neatly into the developing argument in the text.

Boxes in reviews

Boxes are sometimes used where review authors wish to present information on a topic of relevance to their argument, perhaps as background, which is likely to be of interest to only some of the readers (perhaps those from beyond the specific field of the review). Boxes usually contain a substantial amount of text, and may or may not also contain visual elements. Box 12.1 is an example.

The following list summarises the main points included in this chapter in question format, for use as a checklist when manuscripts are nearing completion:

- Is the coverage of the selected content area absolutely up to date? (This question can be especially important when authors have taken some time to respond to referee comments and are about to resubmit their manuscript – a final literature search before submission is recommended.)
- Does the title accurately indicate the content of the paper and effectively attract the attention of the desired readers?
- Does the Abstract/Summary include all information needed to effectively highlight the article's coverage and the contribution it makes to the understanding of the field?
- If the Table of Contents will be published as a separate component of the paper, have the headings been designed so that those seen initially by readers point to all important parts of the article's story?
- Does the Introduction cater effectively for readers who may not be familiar with the finer technical details of the topic or the field?
- Does the article's THM clearly highlight new ideas or a new synthesis of others' work that adds to the understanding of the field?
- Is the THM represented consistently in the four key information sites: title, Abstract, end of Introduction, and Conclusion? (i.e. Do these four parts all tell the same story?)

12.6 Systematic review articles

This chapter provides some guidance on structuring the writing for a systematic review. It does not provide guidance on undertaking such reviews, as that process is a form of research in which previous studies become the "data" for synthesis. However, there are a growing number of excellent resources to guide authors in this endeavour (see e.g. Petticrew & Roberts 2008; Popay et al. 2006; Siddaway et al. 2019). Similarly, the structured nature of systematic reviews requires that the protocol be determined before the review is undertaken, and such protocols can be published (e.g. www.cochranelibrary.com and www.environmentalevidence.org).

A very useful guide to reporting a systematic review has been developed by PRISMA (Preferred Reporting Items for Systematic Reviews and Meta-Analyses; www.prisma-statement.org). The PRISMA statement (Moher et al. 2009) aims to help researchers to report systematic reviews and meta-analyses and provides a 27-item checklist and flow diagram to guide the process.

Writing a systematic review

The title, Abstract, Introduction, Discussion, and Conclusions sections of a systematic review can be written using the guidance provided elsewhere in this book. The only additional thing to keep in mind when writing these sections is that you should be transparent about the fact that the review is a systematic review or meta-analysis (e.g. in the title).

Systematic review manuscripts necessarily follow a specific structure for the Methods and Results sections, with strong emphasis on reporting to maintain transparency in the methods, data, and analysis. This offers confidence in the management of risks of bias in data selection and synthesis, and ultimately in the results of the review. The PRISMA statement sets out, in detail, the expectations for reporting in the Methods and Results sections of a systematic review (Table 12.3).

Table 12.3 Preferred reporting items in the Methods and Results sections for systematic reviews (modified from the PRISMA statement; Moher et al. 2009).

| | Checklist item | |
Section/topic	Methods section	Results section
Protocol and registration	Indicate if a review protocol exists, if and where it can be accessed (e.g. Web address), and, if available, provide registration information including registration number.	
Eligibility criteria	Specify study characteristics (e.g. PICOS,[a] length of follow-up) and report characteristics (e.g. years considered, language, publication status) used as criteria for eligibility, giving rationale.	
Information sources	Describe all information sources (e.g. databases with dates of coverage, contact with study authors to identify additional studies) in the search and date last searched.	
Search	Present full electronic search strategy for at least one database, including any limits used, such that it could be repeated.	
Study selection	State the process for selecting studies (i.e. screening, eligibility, included in systematic review, and, if applicable, included in the meta-analysis).	Give numbers of studies screened, assessed for eligibility, and included in the review, with reasons for exclusions at each stage, ideally with a flow diagram.
Data collection process	Describe method of data extraction from reports (e.g. piloted forms, independently, in duplicate) and any processes for obtaining and confirming data from investigators.	
Data items	List and define all variables for which data were sought (e.g. PICOS, funding sources) and any assumptions and simplifications made.	

Table 12.3 (*Continued*)

93

Writing review articles

Section/topic	Checklist item	
	Methods section	Results section
Study characteristics		For each study, present characteristics for which data were extracted (e.g. study size, PICOS, follow-up period) and provide the citations.
Risk of bias in individual studies	Describe methods used for assessing risk of bias of individual studies (including specification of whether this was done at the study or outcome level), and how this information is to be used in any data synthesis.	Present data on risk of bias of each study and, if available, any outcome level assessment.
Results of individual studies		For all outcomes considered (benefits or harms), present, for each study: (a) simple summary data for each intervention group (b) effect estimates and confidence intervals, ideally with a forest plot.
Summary measures	State the principal summary measures (e.g. risk ratio, difference in means).	
Synthesis of results	Describe the methods of handling data and combining results of studies, if done, including measures of consistency (e.g. I^2) for each meta-analysis.	Present results of each meta-analysis done, including confidence intervals and measures of consistency.
Risk of bias across studies	Specify any assessment of risk of bias that may affect the cumulative evidence (e.g. publication bias, selective reporting within studies).	Present results of any assessment of risk of bias across studies.
Additional analyses	Describe methods of additional analyses (e.g. sensitivity or subgroup analyses, meta-regression), if done, indicating which were pre-specified.	Give results of additional analyses, if done (e.g. sensitivity or subgroup analyses, meta-regression).

[a] PICOS – Population, Intervention, Comparison, Outcome and Study type (Methley et al. 2014)
Source: Modified from the PRISMA statement, Moher et al., 2009.

12.7 Submission and revision of review articles

Submitting a review article

Submitting a review article, whether following invitation by an editor or not, requires similar actions and skills to those discussed for research articles in Chapter 13. The role of the editor and reviewers is to ensure a timely, contemporary, comprehensive, and compelling review and synthesis of a topic of interest to the relevant scientific subdiscipline. Dealings with the journal editor should keep this in mind and aim to encourage acceptance of the review because its synthesis and conclusions offer something new and significant.

Reviews can take time to compile and may be something that a researcher works on as an "additional" activity while undertaking new research or other duties. However, it is important to complete and submit a review in a timely manner to avoid the disappointment of another review or article publishing ahead of you, and to avoid the need to keep returning to the literature for updates if the process drags on. For an invited review, the invitation may be withdrawn by the editor if there is too much delay.

If a review has not been invited or you have not previously sought an editor's interest, use a cover letter to highlight the value of your review (as discussed in Section 13.4). In all cases, ensure that you have followed the Instructions to Contributors and the guidance for manuscript preparation provided throughout this book.

How to respond to editors and referees

Reviews are almost always sent out for peer review, and being invited to write a review or having a positive preliminary response from an editor does not guarantee that a review will be accepted, or accepted in its original form.

The main types of reviewer comments are the same as for a research article; only the "study" and "data" are in a different form. Types of reviewer comments and how to respond to them are covered in Section 14.3 and Table 14.2.

Getting your manuscript published

CHAPTER 13

Submitting a manuscript

Submitting your manuscript to a journal is like entering any competition where success is determined by a group of judges using a defined set of selection criteria. You can optimise your publication success by understanding and meeting the selection criteria of the journal. Many of the criteria related to manuscript preparation may be listed on the journal's website (e.g. Instructions to Contributors and journal scope or aim). Other criteria relate to how a manuscript should conform to the standard of the journal and can only be understood by reading and thinking about the particular journal and understanding the editing and review processes. Here, we describe the editing and review of journal articles and document the main selection criteria used by editors and reviewers. This information will help you to adopt practices that will assist you in developing your publishing strategy and navigating the publishing process, leading to publication success.

13.1 Five practices of successful authors

Success as a scientist is largely measured by the quality and quantity of research output and the impact of that research on other research or practice. Publishing scientific articles is a necessary part of that success. Successful authors adopt five practices to optimise their publication outcomes. They

1 review manuscripts for colleagues and journals and develop a strong framework for research writing and manuscript critique;
2 plan their research and writing to meet the quality assurance criteria that reviewers and editors will impose;
3 carefully select the journal they will submit to and prepare the manuscript content and style to maximise their chances of acceptance;
4 use structured self-review processes and pre-reviews from colleagues to improve the manuscript before submitting it to a journal; and
5 use journal referee reports to improve the manuscript and demonstrate to the journal editor how improvements have been made.

Writing Scientific Research Articles: Strategy and Steps, Third Edition.
Margaret Cargill and Patrick O'Connor.
© 2021 John Wiley & Sons Ltd. Published 2021 by John Wiley & Sons Ltd.

A scientific research article does not produce truth or certainty but documents the observations/measurements, analysis, and interpretation of the authors in the context of previous research. The veracity of findings from a scientific study will be confirmed by subsequent research or application, and may be qualified or amended over time. The peer-review process assists the scientific community in assuring the quality of research before it is published and before it can be examined and used by a wider audience. Peer reviewing is part of the process of turning information into knowledge. The correspondence between the author, reviewer, and editor is part of a collective sensemaking process used to test that new information is worth knowing and acting upon. The system of peer review is not perfect, but it does make a number of critical contributions to the standard of scientific research publications. Specifically, peer review

- confirms that hypotheses have been tested appropriately and that results reported reflect the materials, methods, and analysis tools used;
- confirms that the strength of claims about results and implications are appropriate;
- assists journals to decide whether the focus, novelty, and importance of research are appropriate for their standards, where that is a consideration;
- checks that the presentation and style of content conforms to accepted conventions for production and reader convenience; and
- advises authors and journal editors about how (and, often, where) a manuscript could be improved.

Reviewers are important to the journal editor because they take a critical role in determining the quality of manuscripts. In most cases, they do this as a professional contribution and without payment. Reviewers are important to the author because they bring a critical eye to the content and writing, and highlight how the story can be clarified or more suitably presented. Peer review provides the opportunity to have your ideas, theories, methods, results, analyses, and interpretations considered and commented on by a professional colleague. Responding to the comments of a peer reviewer should be seen as part of the process of testing and legitimising your research results and their meaning.

The best way to develop your understanding of the peer-review process is to carry out peer review yourself. You may be asked to review for journals if you are publishing your own work. If you are not publishing yet, you can offer to review the work of your colleagues or form a journal club and examine the work of already published authors (see Task 13.2). See Chapter 16 for additional ideas on developing reviewing skills.

Traditionally, journals have managed the peer-review process themselves by inviting scientists with reputation in the field of a submitted manuscript to provide a review against their review criteria. Web 2.0 and software advances are facilitating new reviewing models, and possibilities are emerging for manuscript peer review before journal submission or for open Web-based publication of manuscripts with open online post-publication peer review. There is no sign that the importance of peer review of scientific works for publication is diminishing; however, the management and staging of the peer-review processes may change in the future.

13.3 Understanding the editor's role

The editor is responsible for maintaining the reputation of the journal by publishing papers of scientific merit within its scope. Editors use reviewers to assist them in selecting manuscripts and improving them for publication. The editor (or a designated member of the editorial team) will read the manuscript and make the initial decision as to whether it will be sent to reviewers. They will usually reject a manuscript without review if it is outside the scope or aims of the journal, if the language or structure is poor, if it does not include the necessary information, if it has not be prepared according to the Instructions to Contributors, or if there are clear or obvious flaws in the science (see Table 14.1 for a guide to dealing with rejection). A well-prepared manuscript reporting science appropriate to the journal is unlikely to be rejected without review; however, some journal editors may apply a high standard when determining if a manuscript should be reviewed. Journals are emerging which select articles for publication based on scientific and methodological soundness only, and do not apply subjective determination of "impact," "novelty," or "interest" (e.g. PLOS ONE, PeerJ journals).

13.4 The contributor's covering letter

The covering letter you send to the editor with your manuscript (this generally means uploading your message in the appropriate box on the journal's submission website) is an important opportunity to sell your paper. The covering letter allows you to demonstrate that you appreciate the role of the editor and that you have done everything you can to prepare the manuscript to meet the journal's requirements. You can use the covering letter to

- express your belief that the paper is within the scope of the journal and will contribute to meeting its aims;
- state the title of the manuscript and the names of the authors;
- state that the research is new and original and has not been published or submitted elsewhere;
- highlight specific points that reinforce the novelty and significance of the research (avoid simply repeating sentences from the Abstract, which the editor already has to hand);
- highlight any points about the manuscript which may raise questions for the editor (e.g. that a long paper is justified);
- express hope that the presentation is satisfactory; and
- say that you look forward to the reviewers' comments.

Complete Task 13.1 now.

Task 13.1 The contributor's letter as sales pitch

Examine the example covering letter in Figure 13.1 and draw a box around the words which sell the manuscript to the editor most strongly. Check your answers with our suggestions in the Answer pages.

From: A.J. Bond
University of Adelaide, Australia

To: Prof H Zheng
Managing Editor
Journal of Applied Ecology

Date

Dear Prof Zheng,

On behalf of *co-authors* I am pleased to submit the manuscript entitled "Remnant woodland biodiversity gains under 10 years of revealed-price incentive payments." for publication in the *Journal of Applied Ecology*. This study addresses a major challenge of applied ecology; the evaluation of outcomes from "at scale" conservation incentive programs (Payments for Ecosystem Services programs). Our study provides the first example of use of a Before-After-Control-Impact monitoring design to evaluate the effectiveness of incentives for restoration of remnant native vegetation. Additionally, the evaluation of conservation changes is in relation to a program where landholders set their own prices for management and were contracted to undertake management prescriptions over a ten-year period. The study provides evidence that management of remnant native vegetation under a revealed-price incentive program can produce gains in vegetation condition.

We believe the manuscript is very well positioned within the scope of the *Journal of Applied Ecology* and work published in your journal over the past few years. This study provides important evidence for contributors to conservation program effectiveness for managers and policy makers, particularly those working in agricultural landscapes.

The manuscript has been prepared according to the instructions to authors and we hope that the information supplied meets your requirements. No part of this manuscript has been published previously or is under consideration for publication elsewhere. All authors agree to the content of the manuscript.

We thank you in advance for the time it takes to consider our manuscript and we look forward to receiving your response and the feedback of reviewers.

Regards
A.J. Bond

Fig. 13.1 An example covering letter from a manuscript author.

Editors enlist the help of two or more independent researchers to peer review each manuscript and check the quality, novelty, and significance of the work and the presentation of the paper. This work is usually unpaid and is undertaken as part of the professional contribution of researchers to the development of their field of science. Reviewers will

- usually be expert in the general field of the paper (not necessarily in the exact subject);
- almost always have published work in the general field themselves (possibly work that has been cited in the paper);
- be busy with their own research, writing, teaching, administration, family, etc.;
- be willing to review manuscripts but have limited time and patience; and
- have their own preferences and biases about scientific research and writing.

The journal may ask you to nominate potential reviewers at the time of submission, or the editor may choose them from a database or using professional networks. You will not know who the reviewers are. In many but not all cases, depending on the policy of the journal, the reviewers will know the names of the authors. The reviewers will be asked to read the manuscript, write a report about the quality of the work, note any problems, and recommend any changes that would improve it. They will usually be asked to complete an evaluation form about the quality of the manuscript as well, and may also be asked to recommend whether or not the manuscript should be accepted by the journal, or accepted after revisions. The reviewers will return their written report and the evaluation form to the editor, sometimes with annotations on the manuscript (although this is now less common with electronic submission and review).

Journals have their own sets of instructions for reviewers. These are sometimes available on a journal's website, or a colleague who has reviewed for the journal may be able to show them to you. Figure 13.2 provides an example reviewer's evaluation form that includes the main questions to which reviewers are commonly asked to respond. An example of a written report from a reviewer is shown in Figure 13.3. Complete Task 13.2 for practice with these elements.

Referee's Evaluation Form

General questions

1. Does the contribution fall within the scope of the journal? ☐ Yes ☐ No

2. Is the contribution new and original? ☐ Yes ☐ No

3. Is the contribution of sufficiently high impact for the journal? ☐ Yes ☐ No

4. Does the manuscript demonstrate an adequate awareness of relevant research? ☐ Yes ☐ No

5. Are the interpretations and conclusions sound and justified by the data presented? ☐ Yes ☐ No

6. Is the organisation of the manuscript acceptable? ☐ Yes ☐ No

7. Does the title clearly reflect the contents? ☐ Yes ☐ No

8. Is the abstract sufficiently informative, especially when read in isolation? ☐ Yes ☐ No

9. Does the introduction set the manuscript in an international context and show how it builds on previous work? ☐ Yes ☐ No

10. Do the methods and analysis conform to acceptable standards and is sufficient detail presented to comprehend the study? ☐ Yes ☐ No

11. Are all the figures and tables necessary and the captions adequate and informative? ☐ Yes ☐ No

12. Is the length of the manuscript appropriate to the content? ☐ Yes ☐ No

13. Are the references adequate for the subject and length of the manuscript? ☐ Yes ☐ No

14. Is the quality of the English satisfactory? ☐ Yes ☐ No

Please attach a file with any additional comments

Recommendation

☐ Accept without revision

☐ Accept with minor revision

☐ Review again after major revision

☐ Reject

Fig. 13.2 Example manuscript evaluation form showing typical questions for referees to answer.

Task 13.2   Journal club

Form or join a journal club with between 3 and 10 colleagues in related fields of research and arrange to meet regularly (e.g. once a month). Choose some recent articles of interest to the group and arrange to discuss one at each meeting. Each member should use the reviewer evaluation report questions in Figure 13.2 to develop a reviewer's report on the subject article at each meeting. Members should discuss the strengths and weaknesses of the article and any improvements that could have been made. (See Chapter 16 for additional ideas on journal clubs.)

To: Dr S.E. Smith,
Editor, Journal X

Re: Manuscript Number...
Title...
Authors...

Dear Dr Smith,

The paper describes..... . This is a topic which would benefit from additional work such as that described in the manuscript. However, a major concern with the paper is the interpretation and referencing of the literature in the Introduction and Discussion. Related to this is a lack of integration with previous work to explain aspects of the Methods. The paper needs reinterpretation after a thorough investigation of the literature. I recommend that the paper in its current form be rejected but believe that it may be suitable for Journal X after major revision.

Introduction
The Introduction has incorrectly cited [Brown et al. (1981)] who actually showed that......

Methods
Factors relevant to the choice of Methods are: (1) how old were the cultures that were used? 2) Does the age of the culture material affect the results?

Results
The main claim by the authors that their Results showed that...is not correct. Their statement that the results show... needs correction.

Discussion
Relevant references seem to have been overlooked in both the Introduction and Discussion sections, including...

Other queries and suggestions appear as annotations on the attached PDF of the manuscript.

Yours sincerely,

Y.-G. Zhu
Managing Editor

Fig. 13.3 A reviewer's report recommending rejection but noting that the paper would be acceptable with some alterations. Content-specific elements have been deleted.

13.6 Understanding the editor's role (continued)

The editor receives the reports from the reviewers and decides what response will be made to the authors. If the reviewers disagree (especially if there are only two reviewers), the editor will sometimes send the manuscript to another reviewer for an additional opinion. The editor then writes to the corresponding author with the decision that has been made. Responding to these letters from journal editors is a skill in itself and is the subject of Chapter 14.

How to respond to peer reviews

Peer review and journal standards are quality assurance processes which aim to set the publication of scientific research apart from other forms of publishing. Peer review has many benefits for quality assurance but it is not a perfect process. A primary objective is to build trust in published research, both as a foundation for subsequent research and for application based on research findings. Peer review aims to achieve this by acting as a filter, ensuring published research studies have been tested for originality, importance, and validity (within the confines of the study methods). Additionally, the review process seeks to improve the quality of manuscripts by adding the expertise of reviewers to the consideration of the study and its presentation. At its best, peer review is a collegiate conversation seeking win–win outcomes from the clarification of ambiguities and misunderstandings. The main criticism of the process is that is an inherently human one and there is little evidence that it works to improve the quality of published research. Individual reviews can be prone to bias and error, and the review process delays the publication of research and adds to the (unpaid) work of other researchers. The process also potentially stifles innovation, as new ideas must pass the scrutiny of a limited number of reviewers before they can be evaluated by all researchers in a discipline.

The main benefit is that knowledge that manuscripts will be subjected to peer review under transparent standards encourages authors to undertake and write up research to a high standard with the expectation of scrutiny. By striving to meet the standards and rigours of peer review, authors add to and advance research in their field within a framework of integrity and authenticity.

14.1 Rules of thumb for responding to reviews

Critical comments about your research or writing can be difficult to accept and respond to. We recommend the following rules of thumb as a framework for responding to reviewer and editor reports on your manuscript:

- **Rule 1**:It is rare that the author is completely right and the reviewer completely wrong, or that the reviewer is completely right and the author completely wrong.
- **Rule 2**:When responding to a reviewer, the object is to accommodate them by addressing their comments without compromising the message (story) of the paper.

Writing Scientific Research Articles: Strategy and Steps, Third Edition.
Margaret Cargill and Patrick O'Connor.
© 2021 John Wiley & Sons Ltd. Published 2021 by John Wiley & Sons Ltd.

- **Rule 3**:Always show the editor that you are doing everything you can to improve the manuscript.
- **Rule 4**:Rejection and criticism do not automatically mean that the science is not good or that the paper is not well written: consider submitting to other journals, including additional work, or rewriting some or all of the paper.

14.2 How to deal with manuscript rejection

If your manuscript is rejected, it is important to determine the reasons why. The reasons for rejection will inform your decision about how to proceed. Every experienced researcher has a story of rejection, and it can be useful to discuss rejection with a senior colleague to help you see that it is a natural and necessary part of the process of legitimising scientific knowledge. Almost every one of these colleagues will also tell you that all or some of the data from their rejected manuscripts were eventually published. Remember, *everyone* gets rejections. Successful authors are successful at dealing with rejection as well as acceptance. Reasons why your manuscript may have been rejected and recommendations on how to proceed are listed in Table 14.1.

14.3 How to deal with "conditional acceptance" or "revise and resubmit"

Few manuscripts are accepted for publication without some revision. The level of revision varies from minor changes to the language, references, or formatting to major revisions which may require resubmission for fresh reviewing. In fields where journals compete for a share of the new and interesting research in the discipline or a share of the subscription market, journal editors aim to accept high-quality manuscripts as quickly as possible and get them into print (or available online on the journal website) in a timely manner. When the science is obviously interesting and new but the manuscript requires major work before it is acceptable, the editor may reject the paper but encourage rewriting and resubmission. If the manuscript requires some modification but not major restructuring, additional research, or rewriting, the editor may accept it on the condition that recommended changes are made and the article is returned by a set date. An example of a conditional acceptance letter is provided in Figure 14.1. Conditional acceptance gives you the opportunity to consider and incorporate the comments of reviewers and the editor. However, it is not always easy to understand or address reviewer comments.

N.B. The difference between a conditional acceptance and a revise and resubmit response may only be visible in the wording the editor uses in their letter to you. Your first task when you receive the response is to decide what the editor means, and this is not always easy. The editor may use indirect language in the interest of being polite and maintaining your good opinion of the journal for future occasions. If you are in any doubt about the meaning, show the letter to a colleague and discuss it. In any case, you will need to communicate with your co-authors about your response while deciding your strategy for the next stage in the process.

Table 14.1 Reasons for manuscript rejection and recommendations for author response.

Reason for rejection	Response option 1	Response option 2
The content of the paper may not fit the scope of the journal (this could mean it is too specialised, focused on the wrong subject area, or not of enough general interest for the journal's readership). There are clear and obvious flaws in the science.	The editor will have made this decision (usually before the manuscript has been reviewed), and it is usually necessary to revise the manuscript and submit it to a more appropriate journal (check your list of preferred journals; see Table 1.1). The editor may have made this decision before review. Revise the manuscript using the author response guide (Table 14.2) and try to publish the best parts.	If the manuscript has been reviewed, use the referee's reports following rules of thumb 2 and 3 in Section 14.1 and submit to a more appropriate journal (check your list of preferred journals; see Table 1.1).
The language or structure of the manuscript is poor and it could not be sent to referees.	The editor may have made this decision before review. Revise the manuscript using the author response guide (Table 14.2) and resubmit, or submit the manuscript to another journal.	
High-ranking journals need to reject a high proportion of submitted manuscripts even if the reviews are (mostly) positive.	Examine the editor's letter and determine whether there is any encouragement to resubmit a revised manuscript (e.g. "revise and resubmit"). If there is, revise and resubmit following rules of thumb 2 and 3 in Section 14.1.	If resubmission is not encouraged, revise the manuscript using the referee's reports following rules of thumb 2 and 3 in Section 14.1 and submit to the next journal on your list (see Table 1.1).
Referees may not have read or understood the paper thoroughly enough to appreciate it. Recommendations from the referee to the editor may not be clear or may have been misinterpreted. Something may have annoyed the referee: they are unpredictable and can be helpful or (sometimes) unhelpful.	You can appeal to the editor, although this is unlikely to be successful unless a major error of judgement has been made by the referee or the editor. It is always wise to make the uncontroversial changes recommended before appealing, resubmitting, or submitting to a different journal.	Revise and resubmit, or submit to a new journal. Clarify any issues which have caused problems by revising the text. If resubmitting to the same journal, make note in your letter to the editor of any misunderstanding, any supportive comments from referees, and the improvements you have made to the manuscript.

From: Prof Y-G Zhu,
Managing Editor

Journal X

Dear Prof S.E. Smith,

Thank you for submitting your revised manuscript to Journal X. I have now received the reviewers' reports a recommendation from the Associate Editor who handled the review process. Copies of the reports are attached. As you will see, the reviewers are positive about the value of the work although they have made a number of suggestions for further improvement. I have considered your paper in light of the comments received and I would like to invite you to prepare a revision.

The referees and Associate Editor are all impressed with your efforts thus far. However, they would like to see more work on the presentation, especially focusing on the content of the discussion. I draw your attention to the main points from two referees of your paper.

Referee 1:

- The Methods section does not give sufficient information, particularly about the sampling methods used.
- The results in Tables 1 and 2 are closely related and can be combined into a single table.
- The conclusion that there is a strong positive correlation between the number of organisms and soil salinity needs a stronger statistical basis.
- The results in Figure 3 are very preliminary-this really requires another survey. If this is not possible, the Figure should be deleted.

Referee 2:

- There are inadequacies in the Methods section, as indicated on the typescript.
- The Discussion is not well focussed and does not include some important relevant publications, e.g. Cavagnaro et al. (2018) and Bond et al (2019).
- The conclusion is interesting but can be greatly strengthened. In particular, the findings are different from those of Loch et al. (2017), a study done in the USA. The work in your paper is in fact the first study of its kind outside Europe and North America and this should be highlighted.

I would be grateful if you would consider the above comments and those made in the attached reports, and revise the paper to take account of the salient points. Please note that Journal X does not automatically accept papers after revision, and an invitation to revise a manuscript does not represent commitment to eventual publication on our part. We will reject revised manuscripts if they are returned without satisfactory responses to the reviewers' comments. When returning the revised paper, please show point-by-point how you have dealt with the various comments in the appropriate section of the submission form.

Please return your revision within four weeks of the date of this message. If you need longer, please let us know so we can update our system accordingly.
We look forward to hearing from you in due course.

Yours sincerely,

Y.-G. Zhu
Managing Editor

Fig. 14.1 An adapted example of a conditional acceptance letter from a journal editor. (N.B. Reviewer reports would be attached to this letter.)

There are many ways to deal with reviewers' comments and you will develop your own strategies. Here, we outline an approach used by many experienced authors:

- Do not get angry or offended by the comments. The reviewer or editor may have misunderstood something or you may have communicated it poorly. Dealing with reviewers' comments is part of the publishing process and comments should not be seen as a personal attack on your credibility as a scientist or writer.

- Read the comments and check the manuscript to make sure you understand what the referee or editor is asking you to do.
- Group similar comments together from multiple reviews so that they can be addressed collectively.
- Highlight any comments which are difficult to respond to or are unclear.
- Show the difficult comments to co-authors or colleagues and seek their advice about how to deal with them. If comments are still difficult or unclear, or they annoy you, leave them for a few days (not more than a week) and return to address them when you have had time to absorb them.
- Review the rules of thumb (Section 14.1).
- Make all the small changes which do not require major rewriting and note each change in a letter to the editor.
- Respond to any major comments using the suggested responses in Table 14.2.
- Prepare your responses in the format demonstrated in Figure 14.2.

Main types of reviewer comments

Every review is different and will present different challenges to which to respond. However, the majority of reviewer comments fall into eight categories:

1 The aims of the study are not clear.
2 The theoretical premise or "school of thought" on which the work is based is challenged.
3 The experimental design or analysis methods are challenged.
4 You are asked to supply additional data or information that would improve the paper.
5 You are asked to present results differently.
6 You are asked to remove information or discussion.
7 The conclusions are incorrect, weak, or too strong.
8 The referee has unspecific negative comments, e.g. "poorly designed," "poorly written," "badly organised," "tables are too large," "relevant literature not cited," or "English is poor."

Decide which of these categories each of the difficult comments falls into. If these categories do not cover a comment you have received, decide what it is really about and consider whether the approaches recommended in Table 14.2 are still appropriate. If you have a conditional acceptance from the editor, then none of the comments is likely to be enough to stop you publishing the paper. In any situation where you need to revise your manuscript in the light of referee comments, the main exercise is to maintain the integrity of your story while accommodating the referees and editor.

Table 14.2 sets out each of the main types of comments you are likely to receive from the referees and editor and recommends a series of approaches to responding to them. The recommended responses to each comment type range from easy to more difficult, and some comment types may require a mixture of responses. Many reviewers' comments can be addressed by appropriate use of the two most powerful tools available to writers of scientific articles:

- **Citing the published literature**: Published works have already been reviewed and accepted by the scientific community. The findings and conclusions that

Table 14.2 Author response guide on how and where (in the manuscript and in correspondence with the editor) to deal with referee reports.

Comment type	Author response	Location in manuscript	Cross-reference within this book
The aims or novelty of the study are not clear	Rewrite the aims to state them clearly (write as a hypothesis if the reviewer insists or if it is a requirement or convention of the journal, or make the case not to if it is not a convention)	Introduction (Stage 4)	Section 8.6
	Ensure the aims are consistent with the experimental design	Cross-check Introduction (Stage 4) with Methods	Chapters 7 and 8
	Ensure the research gap or problem to be addressed is clear and emphasised	Introduction (Stages 3 and 2; Stage 4 may need rewriting)	Chapter 8
	Ensure the discussion refers back to the aims	Cross-check Discussion with Introduction (Stage 4)	Sections 8.6 and 9.1
The theoretical premise or "school of thought" on which the work is based is challenged	Ensure you have shown the diversity of theories (citing the literature) and explain why the one you are testing is valid or superior in the context of the study	Introduction (Stages 2 and 4; may be re-emphasised in the Discussion	Chapter 8 and Section 9.1
	Include caveats (conditions on when the results or theory may not apply)	Discussion	Chapter 9
The experimental design or analysis methods are challenged	Defend the design or analysis on its merits	In letter responding to referee's comments	Section 14.3
	Refer to previously published examples using the design or analysis (citing the literature)	Methods – also in the response to referee's comments	Chapter 7 and Section 14.3
	Include additional information on the design or analysis, if available	Methods (may be re-emphasised in the Discussion)	Chapters 7 and 9
You are asked to supply additional data or information that would improve the paper	Supply the additional data if you can	Results (may include other sections)	Chapters 5 and 6
	Explain why the requested information is outside the scope of the study, that the study is coherent and advances the field without the addition, or that there is a need for future research (and offer direction for this research, if appropriate)	In letter responding to referee's comments	Section 14.3
	If major changes are required, consider rewriting the paper to make additional information unnecessary	All sections	Chapter 4

Situation	Action	Where	Reference
You are asked to present results differently	Ensure the suggested changes will improve the presentation and support the story of the paper. Make the changes where possible, or add to supplementary material	Results	Chapter 5
You are asked to remove information or discussion	Remove the information if you can do so without changing the "story." Move some results to the supplementary material. You can ask a colleague to make suggestions on where to make cuts if the referee is not clear	Where indicated (usually Introduction and Discussion)	Various
	If cuts would alter your story too much, weigh up the positives in the referee reports and make your case to the editor for retaining information	In letter responding to referee's comments	Section 14.3
The conclusions are incorrect, weak, or too strong	Ensure the discussion is tied to the aims at the beginning of the paper	Cross-check Discussion with Introduction (Stage 4)	Chapter 9 and Section 8.6
	Reassess the literature you have cited and make a case to the editor if there is adequate supporting literature (check and cite supporting literature)	Discussion and in letter responding to referee's comments	Chapter 9 and Section 14.3
	Ensure that all your statements are justified and the strength is appropriate	Discussion	Section 9.2
	Include constraints (conditions on when the results may not apply)	Discussion	Chapter 9
The referee has unspecific negative comments (e.g. poorly designed, written, or organised)	Discuss the referees' comments with a peer	All sections	Various
	Restate or rewrite the section(s) and note each change you make to the editor	In letter responding to referee's comments	Section 14.3
	Point out to the editor all the work you have done to improve the paper (i.e. build up a body of positives: "I have addressed point 1 by. . .")	In letter responding to referee's comments	Section 14.3
	Seek language or editorial assistance if the writing or grammar is criticised	Relevant sections	Chapter 17

To: Prof Y.-G. Zhu,
Editor, Journal X

Re: Manuscript Number X123/2020
[Title]
[Author names]
[Date]

Dear Prof Zhu,

Thank you for reviewing the manuscript and the conditional acceptance pending revisions.
Thanks also to the Associate Editor and referees for their positive and constructive comments on
the manuscript. We have addressed all the comments of the two reviewers in the attached response to
reviewers, including the key points you referred us to. We have also attended to the formatting and
language of the manuscript according to your suggestions.

We believe the manuscript tis greatly improved and are confident it is now acceptable for publication. We
look forward to receiving your response to the changes we have made, and if you require any further
information please do not hesitate to contact me.

Yours sincerely,
Prof S.E. Smith

Fig.14.2 Example of an adapted author response letter to an editor, responding to reviewers' comments. (NB: Specific responses would be attached to this letter as per Table 14.3.)

have been published by different authors can be compared and contrasted and used to develop an idea or support an argument.

- **Improving the structure of the manuscript**: The structure and logic of each section and subsection of a scientific article are described in this book. Reviewing the relevant chapters will help you to deal with reviewers' comments by allowing you to improve the structure of your ideas or arguments.

Use Table 14.2 to decide on the appropriate response(s) to comments and the place(s) in the manuscript where changes should be made (the reviewers' comments may indicate this). The table also indicates which sections of this book to review as you deal with reviewers' comments.

Complete Task 14.1 to consolidate your understanding.

Task 14.1 Analysing an authentic example

Ask a colleague who has had an article reviewed to show you the reviewers' comments and their responses.

1 Decide which of the eight types of reviewer comments listed in Table 14.2 were made.
2 Check whether the responses the author made fit the suggested response types in Table 14.2.
3 Discuss the thinking behind the responses with the author.

See Chapter 16 for additional suggestions about using previous reviews as a training tool.

It is important to respond quickly to reviewers' comments and the editor's recommendation about publishing the manuscript. This is true regardless of whether the manuscript has been accepted with minor changes or you have been encouraged to resubmit it after making major revisions. As with the covering letter you sent when you originally submitted the manuscript, the letter accompanying the revised manuscript is an opportunity to demonstrate that you appreciate the role of the editor and that you have done everything you can to improve the manuscript in order to meet the journal's and the reviewers' requirements. Use the letter responding to reviewers' comments and a table of responses to do the following:

- Advise the editor you have made the changes requested, including changes to language, formatting, and supplementary material, and that the article is now ready for reconsideration (e.g. Figure 14.2).
- Prepare a table with columns, rows, and headings as per the example in Table 14.3.
- In the table, list each of the comments from the reviewers in one column, your response to each in the next column, and the changes made in the manuscript in the final column, including section and line numbers. This way, the editor can see each comment and how it has been addressed, and the changes made in the manuscript.

In responding to reviewers' comments, you can point out supportive comments by the referees and any disagreements between them if it assists in making clear why you have taken the approach you have. Defend your work if a referee is factually wrong (this is another chance to cite key published papers supporting your argument). Recheck that the changes to the manuscript conform to the guidelines in the Instructions to Contributors (e.g. formatting, length, style).

Send the revised manuscript back to the editor, together with your letter and table of responses to reviews.

Table 14.3 Example author responses to reviewer comments.

Item	Reviewer 1 comments	Authors' response	Revisions made
1.	The paper is mostly clearly written. The introduction well describes the weaknesses of previous studies and advocates for the need of sound sampling design over long time periods. The sampling design described in the methods is a strength of the study, the methods of analysis are sound as well.	Thank you for your positive comments. Your comments and suggestions have helped us to make valuable improvements to the manuscript.	Revisions made addressing the concerns raised are described in detail below.
2.	Results are not clearly presented, especially in the Abstract this part is confusing for the reader. I think results should be better synthesised here and presented in a clearer way.	Thank you for the comments. We have revised the relevant text in the Abstract.	Revised text in Abstract: "3. Modest improvements were detected in native plant species richness, log abundance and grazing pressure at impact sites compared with background changes observed at control sites."
3.	Delete: "despite the large investments being made." as it's redundant.	Thank you for the comment, we have deleted this text as suggested.	Text deleted.

Item	Reviewer 2 comments	Authors' response	Revisions made
4.	Cite the official listing of threatened ecological communities.	Thank you for the comment. We have added the citation as requested.	Citation added.
5.	Fig 1 map approximate locations, make this clear somewhere that some of the data is off-limits, such as specific locations. Reviewer 1 No legend for yellow colour in MAP.	Thank you for the comments. The map and figure caption have been updated as indicated.	Added note to figure 1 caption: "NB Study site location accuracy has been reduced to protect the privacy of program participants." Updated map includes legend for yellow colour as "Other land cover."
6.	Where assessments done within the same plot?	Thank you for the comment. We have clarified this by inserting text in the Methods section.	Text inserted in Methods section: "A set of variables relating to habitat condition were measured at a sample plot within each site, with the same sample plot measured in both assessments."
7.	Should I be imagining a line between the vector labels and the origin? Figure 3b	Thank you for the comment. We've revised the figure to show arrows for the vectors.	Revised Figure 3.
#	Etc.		

CHAPTER 15

A process for preparing a manuscript

There are many different ways to proceed towards preparing a manuscript for submission to a journal, but the process often seems to take a very long time and to involve a considerable amount of back-tracking and reworking. Indeed, multiple drafts are a necessary part of manuscript writing – as co-authors make their respective contributions and the paper's story is refined and strengthened – but it is in everyone's interests to streamline the process as much as possible. Here, we present a suggested approach to manuscript mapping through a series of steps which aim to develop the story and main components of the different sections of the manuscript in consultation with co-authors.

Before and alongside the mapping and writing process, you should take additional steps to clarify, develop, and target the story of your manuscript and plan the writing:

- Who is the target audience for the paper, how significant is the story told by the data, and thus which journal should be selected as the target?
- How will the work of preparing the manuscript be divided up (i.e. who will do what)?
- Who will be listed as authors, and in what order will their names be shown? Who should be acknowledged for assistance? (It might be helpful to consult a source such as the website of the International Committee of Medical Journal Editors (www.icmje.org) for criteria to use in determining who qualifies as an author)
- What timeline is feasible? At which stages will the co-authors read drafts? (Once a decision has been made, you can insert steps at relevant places in the manuscript mapping in Section 15.1 to allow for feedback from co-authors. A clear, documented process (e.g. a Gantt Chart) for developing the manuscript can help avoid confusion and frustration and ensure co-authors agree to appropriately prioritise their contribution to the manuscript.)

It can help to give a short talk to a small group of your colleagues and present some background and reasons for the research (Stages 2 and 3 of the Introduction); the aims or hypothesis; an outline of key methods; all the data needed to tell the story (all the figures, tables, and other text); and a discussion of the results and their meaning. Ask the group to provide feedback on anything which was not clear in your presentation and to ask any questions they have about the research.

Writing Scientific Research Articles: Strategy and Steps, Third Edition.
Margaret Cargill and Patrick O'Connor.
© 2021 John Wiley & Sons Ltd. Published 2021 by John Wiley & Sons Ltd.

15.1 Manuscript mapping

This section describes steps for outlining or "mapping out" your manuscript with co-authors using what you have learned about the key components of each section of a manuscript for publication in an international journal. The process aims to build the skeleton of your manuscript starting from the results bundles. Developing the story and components of your manuscript in this way (with co-authors) can save a great deal of time in revisions.

Step 1: Manuscript outline (with co-authors)

Present the main result bundles (figures, tables and/or dot point text) in an ordered manner for discussion with co-authors. Make dot points for each results group and indicate the section of the manuscript where they should be placed and developed (e.g. dot points under a figure indicating results and elements of discussion):

1 key elements of evidence/results (Results);
2 comments on results (Results);
3 relationship to study hypothesis/purpose, including agreement/disagreement (Introduction/Discussion);
4 explanation of/speculation on results (Discussion);
5 (including) contrast or discussion in relation to key citations (Discussion);
6 limitations to generalisability (D); may include points to add in Methods (Methods) or model selection (Introduction);
7 implications/meaning (Discussion/Introduction); and
8 potential recommendation or application (Discussion/Conclusion).

Organise the dot points for inclusion in drafting of the appropriate sections.

Step 2: Take-home message development

Don't write the manuscript, write the main outcome:

• describe what your results "say" (synthesis of Points 1 and 2);
• describe what your results mean in their context (What is the significance of these results? Combine from Points 4 and 5);
• describe who needs to know this (= audience for the paper); and
• describe why they need to know (What contribution do these results make to the field? What would your selected audience miss out on if they never read your paper?).

Synthesise the take-home message (THM) in 25 words or less. There may be multiple results to combine into a single THM. Ensure your co-authors endorse the THM and centre your manuscript drafting around it.

Step 3: Title draft

Based on your THM, write down some options for a working title (two or three alternatives).

Decide which main result groups will go in the manuscript – do not include more than **six** (tables and figures) without strong justification. For each individual item, check it contains only information essential to supporting the THM. Decide if some items should go into Appendices/Supplementary Information (leaving six or fewer for the manuscript). Write the legends for each so that they stand alone without the text.

Step 5: Results and Discussion writing

Write the first draft of the Results (Chapter 6) and Discussion/Conclusion (Chapter 9) from the dot points collected in Step 1 by producing

1 section headings;
2 topic sentences; and
3 paragraphs.

Circulate for feedback. Continue until there is agreement.

Step 6: Remaining sections

Write the remaining sections:

1 the Introduction (incorporating dot points from Step 1) – consider writing the stages in the order 4, 3, 1, 2, with Stage 5 (if present) at the appropriate place for your particular story and Stage 6 at the end, if required (see Chapter 8 for details);
2 the Abstract (Chapter 11);
3 the Methods (or equivalent) section, to fit the results presented and discussed (Chapter 7); and
4 a set of keywords (Chapter 10).

Review the document with co-authors, prepare for publication, and submit.
To refine your manuscript, follow the suggestions in Section 15.2.

15.2 Editing procedures

1 Put the completed draft aside for a while. The literature on this topic suggests that you need at least 48 hours away from a draft before you can read what you actually wrote, as opposed to what you think you wrote.
2 When you come back to the document, print off a paper copy and read it through from the beginning with the aim of identifying places where content changes are needed. Do not stop to make any changes; just put marks in the margin or under problem words to indicate the places you will need to return to later.
3 Once you have reached the end of the document, go back to the beginning. Work on improving each problem you identified.

4 Edit it again, as before.

5 Do this as many times as necessary. When you have completed this part of the process, you should be satisfied with the science of what you have written.

6 Now edit for so-called discourse features: these are the language features that contribute to the flow and linking of the sections and sentences.

- Check that informative subheadings appear wherever they are needed.
- Check that paragraphs have topic sentences where appropriate.
- Check that paragraphs and sentences follow our guidelines on leading from the general to the particular and on giving old information before new (see Chapter 8).

7 Edit for spelling, punctuation, and grammar.

- Check especially for the mistakes you often make – use the Find feature of your word processor.
- *Always* have the computer's spelling checker switched on, but remember its limitations and that it cannot identify where you have used a word that is correctly spelled but is not the correct one in the context (e.g. if you type *there* where you mean *their*, or *it's* where you mean *its*). You will also need to add to the program's dictionary all the technical terms you use (checking carefully that they are spelled correctly when you add them). Then you can be confident that every time a red wiggly line appears, there really is an issue to be addressed.
- Check for punctuation and italics, especially *et al.* and species names. (Different journals have different conventions about these issues, so make sure you check in the Instructions to Contributors to find out what applies in the journal where you will submit.)
- If you use English as an additional language (EAL), editing your own writing for grammatical accuracy needs special care. We suggest that you use a ruler and hard copy of the text (i.e. do not try to do this on the computer screen). Start with the *last* sentence of a section and lay the ruler under the last line. Read the sentence and check its grammar, making sure that the verbs and subjects agree, that singular and plural forms are used appropriately, that the verb tense is correct, and that articles (a/an/the) are used appropriately. Then move the ruler up the page and read the sentence before the one you just checked. In this way, you are less likely to be distracted by issues other than the ones you are supposed to be looking for: the grammatical ones. Remember, you are already happy with the science of the manuscript, after completing Steps 1–5 as many times as necessary! Now you are only focusing on the grammar.

8 Edit for the correctness and consistency of the referencing and the reference list.

- If you are using one of the commercially available bibliographic software programs, such as Endnote or Reference Manager, most of this step has been done for you, but you will still need to check that the output of the program appears as you want it and that no entries have been produced that have anomalies or inconsistencies, which can occur if data has been entered into the program incorrectly.
- If you have produced the reference list manually, you will need to check carefully for the following three things:
  (i) Does every reference in the text have a corresponding entry in the list?
  (ii) Does every entry in the list appear at least once in the text?
  (iii) Do all references in the text and all entries in the list follow the style stipulated by the journal exactly (i.e. including punctuation, spacing, use of italic and bold fonts, and capitalisation)?

Table 15.1 Checklist for review of paper drafts. 119

Criterion	Reviewer's comments
1 Does the title reflect accurately the content of the paper?	
2 Are the significant words in the title near the beginning to catch a reader's attention?	
3 Does the Introduction begin with the big issue of topical/scientific interest and then narrow down to the specific topic of the paper?	
4 Does the Introduction locate the study effectively within the recent international literature in the field?	
5 Does the Introduction highlight a gap that the research fills, or present a need to extend knowledge in a particular area? (Does it say why the work was done?)	
6 Does the Introduction end with a clear statement of the aim/hypothesis of the research, or summarise the main activity or findings of the paper (depending on the field and relevant journal conventions)?	
7 Are the methods, including statistical analysis, appropriate for the questions addressed and the study conducted?	
8 Are the materials and methods given in enough detail to convince a reader of the credibility of the results?	
9 Do the results provide answers to the questions raised in the Introduction, or fulfil the objectives given?	
10 Are the results presented in a logical order (similar to the order in which either the aims and methods or the Discussion is presented)?	
11 Are all the tables and figures necessary to tell the story of the paper? Could any be combined or deleted?	
12 Do all the tables and figures stand alone? (i.e. Can readers understand them without going back to read the text of the paper?)	
13 Does the Discussion (if in a separate section) begin with a reference to the original aim/hypothesis/question?	
14 Are the results compared with other relevant findings from the literature? Are you aware of any other comparisons that could be made? Are appropriate explanations/speculations included about reasons for observed similarities, differences, and other outcomes?	
15 Are appropriate statements made about the wider significance of the results, their limitations, or their implications for practice and future research directions?	
16 Does the paper end with an appropriate concluding paragraph or section that emphasises the key message(s) and their significance to the field?	
17 Is the list of references complete (all the works in the list are referred to in the paper, and all the works referred to in the paper are in the list)?	
18 Are the reference list and in-text references formatted accurately and in the right style for the target journal?	
19 Does the Abstract/Summary include all the information required by the journal, and does it highlight appropriately the key results and their significance?	
20 Does the Abstract/Summary adhere to the word limit and follow the prescribed format of the target journal?	
21 Are the selected keywords those that will best allow the article to be located by the full range of its prospective readers?	
22 What additional comments do you have for strengthening the paper?	

9 Edit for layout. View each page singly using Print Preview to ensure that headings stay with the following text and running headers appear or not as stipulated in the Instructions to Contributors for your target journal.

10 Check that you have followed the formatting requirements as provided in the Instructions to Contributors, including in regard to the placement in the manuscript or in separate files of tables and figures and their titles and legends, and the provision of any supplementary data to appear on an associated website, if applicable.

11 Final check. Do a final read-through to catch the "little" mistakes that may have slipped by. It can be very helpful to ask a colleague or friend to do this for you – remember also to make yourself available to do the same for them when their turn comes to submit a manuscript.

15.3 A pre-review checklist

Now you are ready to ask for some serious feedback on the article from people outside the author team. One option for this step is to provide your critical reader with a list of questions to respond to. In Table 15.1, we provide such a list, which has been developed on the basis of the material covered in this book. Another option, perhaps to be used after the checklist, is to ask an experienced colleague to pre-review your manuscript; that is, to read it as if they were reviewing it for the journal. If appropriate, you could provide them with the example referee's evaluation form given in Figure 13.2.

Once you have responded to the feedback received in this way and done a final check, you are ready to submit your manuscript. Good luck!

Developing your writing
and publication skills further

CHAPTER 16

Skill-development strategies for groups and individuals

A number of effective strategies and activities can be implemented within research groups, laboratories, or departments to provide a structure or focus for developing publication skills and capacity. At one end of the spectrum, these can be organised by the senior scientists, with students and junior members encouraged or required to participate. At the other end, groups of students or early-career researchers can join together to set up activities they think will benefit their own development, and request input from the senior staff as appropriate.

If your group is located in a country where English is not the working language, then the extent to which these activities take place in English must be decided on a case-by-case basis. It can be helpful to involve at the planning stage an English-teaching professional with relevant expertise to discuss where and how English improvement can be built into the activities. Many of the sections of this book are candidate materials for structured input to such sessions, perhaps to be followed by a time for a discussion of someone's draft paper, or of the slides for an upcoming conference presentation. A more detailed introduction to the teaching methodology used in this book is provided in Burgess & Cargill (2013).

The following sections present some ideas for different types of activities that can be used. We recommend that any strategy be planned to have a limited duration (e.g. meeting every 2 weeks for 3 months, followed by a review), an agreed set of objectives, and explicit ground rules for how sessions will run, preferably agreed upon by all participants at the first meeting.

16.1 Journal clubs

A journal club is a popular strategy used in many science fields to build levels of knowledge in specific areas. It involves all members of the club reading the same journal article (nominated by the group leader or a designated group member) and then coming together to discuss it in depth. The discussion sessions are chaired by a member of the group (this role usually rotates among the membership), who is often expected to identify particular points within the article for focused discussion.

An additional component can be added to the end of these sessions to include a publication skill emphasis. Participants can be asked to analyse one of more of the

Writing Scientific Research Articles: Strategy and Steps, Third Edition.
Margaret Cargill and Patrick O'Connor.
© 2021 John Wiley & Sons Ltd. Published 2021 by John Wiley & Sons Ltd.

article's sections (its title, Abstract, Discussion, etc.) using tasks from the relevant sections of this book. The aim would be to answer questions such as these:

- Is this section effective in terms of communicating this content to its intended audience?
- What makes it effective in your opinion?
- Can you find examples of the techniques highlighted in this book that contribute to the effectiveness?
- Can you identify additional features that make it effective?
- Can you identify anything that could be improved?
- Can you identify places in the text where the authors have highlighted aspects of the research that make it relevant to the journal in which it is published? (Consider both journal scope and standard here.)

16.2 Writing groups

"Writing group" is a general name for any group of people who come together on a regular basis to enhance their progress on individual writing projects: in this context, probably article manuscripts or thesis chapters. Writing groups may be facilitated (a more experienced person provides leadership or input) or unfacilitated (the group members run the group activities themselves). Either approach can be useful, depending on the circumstances, work patterns, and learning-style preferences of the participants.

At a basic level, two or three people can commit to meet on a regular basis to read one another's drafts, with an agreement made at the end of each meeting about who will provide a draft section to the others by an agreed date, for discussion at the next meeting.

16.3 Selecting feedback strategies for different purposes

You may be asked to give feedback on another person's writing in the context of a writing group, as a personal request, or in a more formal capacity as a reviewer for a conference or a journal. As it is rare for training to be provided on the giving of feedback, we present here some comments for you to consider as you approach the task.

Before you give feedback on someone else's writing, it is helpful to clarify the role you have been asked to play. Writers often have a strong emotional investment in their work, and they can sometimes feel under personal attack if they receive comments on it that do not fit with their views of the relationship between writer and reviewer and the role they expect the reviewer to take. So, when someone asks for your feedback, it can be helpful to discuss with them what type they are seeking and what role they want you to take in this particular instance.

One possible feedback type that can be requested is "just the content" or "just the science," with the requester not expecting comments on the language used to express that content. This request is very difficult to carry out for many reviewers; one way to do it is to use the checklist for review of paper drafts (Table 15.1) without annotating the draft itself at all. Alternatively, the writer might seek feedback on the main points of the content (the take-home message (THM) or story of the paper) before they undertake the writing of the full manuscript. This can involve reviewing the answers to the four key questions given in Task 4.1, plus the full set of tables and figures that provide the evidence for the story.

Once feedback is being sought on a full draft of a paper, it is most likely that comments will be forthcoming on all aspects of the text. In this situation, it is useful for the reviewer to think about what role they will adopt, perhaps by reflecting on the following questions:

- To what extent is my purpose to coach (encourage and suggest ways of improving in a supportive way)?
- To what extent is my purpose to act as a gatekeeper (decide whether the work is good enough or appropriate for its purpose)?
- To what extent is my purpose to teach (focus on helping the writer learn things that will become part of their repertoire of skills for the future)?
- What other purposes do I have?

Once you have made some decisions about these points, it may be helpful to think about how much power you want to adopt in your relationship to the writer:

- Do you want to appear as an expert who definitely knows the answers and whose advice must be followed?
- Do you want to appear as a more experienced colleague who can suggest things on the basis of your experience and whose advice should be seriously considered?
- Do you want to appear as a fellow struggler with the issues, someone who is also learning how to write for the international English-language literature, who can act as an example of the intended audience, and who can apply the learning from this book to make suggestions and see if the writer agrees with them?
- Do you want to blend these approaches, adopting more of one in some areas (e.g. the science) and more of another in others (e.g. the language)?

Your answers to these questions may change the words you use to provide written feedback on a draft. For example, in what circumstances would you be more likely to use each of the following options?

- "More explanation needed."
- "Not sure what you mean here."
- "Move this to the Introduction."
- "This may fit better in the Introduction."

In thinking more globally about your feedback style, it can be useful to consider which of the following feedback strategies you have used before, and which you would like to try in the future:

- Commenting on aspects that have been well done before pointing out things to be improved.
- Using different coloured inks for different categories of feedback (e.g. science and language).
- Restricting yourself to the most important issues; intentionally not correcting everything in the case of early drafts.

- Providing a summary at the end of the document of both the positive aspects and the most important changes you recommend.
- Recommending other sources of help: other people to talk to, books or electronic resources to consult.
- Using a set of symbols to indicate the types of issues needing attention, instead of or as well as writing proposed corrections on the manuscript; for example:

sp	=	spelling
p	=	punctuation
sing/pl	=	wrong choice of singular or plural form
wo	=	word order
agt	=	agreement between subject and verb
t	=	tense
art	=	article (*a/an*, *the*, or no article)
obn	=	put old information before new information

It is likely that your answers to all the questions listed here will depend on an even broader set of factors:

- your seniority (how much experience you have);
- your institutional role (what your job requires you to do);
- your personality;
- your relationship with the writer; and
- what the writer has asked you to do.

Finding an appropriate balance in a given situation between all the possible ways of responding can be a challenge, but progress towards this goal can be extremely rewarding. In the end, it contributes to a skill-set that is of considerable importance in the work of a publishing researcher: the ability to give feedback that is rigorous, constructive, and inclusive. A more detailed discussion of these issues is provided in Cadman & Cargill (2007).

16.4 Becoming a reviewer

The discussion in Section 16.3 opens up many of the issues that are relevant to the progression of a young scientist from writer only to reviewer for journals or conferences; however, one important difference from the situations outlined there is that the opportunity for direct discussion with the author of a manuscript is removed once feedback is being provided in a formal review context. The reviewer's primary responsibility now is to the journal editor, to assist them in making a decision about the article's suitability for publication in the journal they represent. Nevertheless, reviewers are asked to provide constructive feedback to authors in terms of improvements needed to help make an article suitable for publication. It is here that the principles and strategies introduced in Section 16.3 may usefully apply, particularly as submitted reviews are often sent to authors unaltered.

Training for writing reviews for journals (or conferences) is most often informal; for example, academic supervisors may involve their graduate students in the process of producing a review (with the permission of the journal editor). Recommendations for more systematic training have recently begun to appear. Wiley provides reviewer resources and links to a wide range of additional sources of advice and training online (https://authorservices.wiley.com/Reviewers/journal-reviewers/tools-and-resources/index.html). Another set of training materials that may be of interest is offered by the *British Medical Journal* (*BMJ*) (https://www.bmj.com/about-bmj/resources-reviewers/training-materials). The format used there could serve as a basis for developing journal-specific or field-specific training in other contexts.

16.5 Training for responding to reviewers

To move beyond the necessarily general advice on this topic provided earlier in the book, we can suggest the following training strategy. It requires that one published member of your research group be willing to share with others the documents that represent the full process of getting one of their articles accepted for publication. In our experience, this is usually a more senior member of the group who has an interest in developing the capacity of less experienced members. A suggested process for a training workshop (or a series of workshop meetings) runs as follows:

1 The author of the paper provides to each workshop participant copies of the originally submitted manuscript and the journal's initial response to it: the editor's letter and the referee reports.
2 Participants are asked to read these documents thoroughly. They then form small groups and discuss how they would have responded to the editor's and referees' comments. This step can include drafting sentences to include in the response letter to the editor and drafting changes to the manuscript itself.
3 Each small group shares their proposed responses with the large group. The author then describes what was actually done in response and distributes copies of the written response that was sent to the journal. It is helpful if the author includes here a description of the emotional response to the editor's letter that was felt by the corresponding author and the other members of the author team, and how those feelings were dealt with.
4 The small groups re-form. Participants read the response document, identify differences between it and their first ideas, and discuss possible reasons for them.
5 The large group reconvenes, and the author comments on the issues identified by each small group in Step 4.
6 If there was a second round of reviewing, the process can be repeated if there are new insights to be gained from doing so. Otherwise, the author can just explain the final outcome.
7 Participants are asked to summarise what they have learned from the workshop, in terms of both strategies for preparing their own responses and points to pay attention to in the original writing and editing of the manuscript prior to submission.
8 In English as an additional language (EAL) contexts, it is useful if participants also take note of any useful sentences or expressions from the example responses discussed in Steps 4 and 6 that could be re-used in their own writing.

CHAPTER 17

Developing discipline-specific English skills

It can be helpful to think of the English you need to write about your research as one English among many Englishes: the English of marine biology, for example, or the English of plant biotechnology. Therefore, to a certain extent, people new to a research field need to develop their discipline-specific English even if English is their first language. We have included the aspects of English usage that are of general interest for scientist authors in the previous chapters on writing each section of an article. This chapter, on the other hand, focuses on those aspects of English grammar and usage that are of particular relevance to science authors who use English as an additional language (EAL). We begin with a discussion of the expectations of editors and reviewers, and their implications for authors. We then introduce two strategies that can be useful for developing discipline-specific English writing skills: the concept of sentence templates, and a family of computer-based tools called concordancers. Finally, we focus on a selected range of features of scientific writing in English that we find present problems for many EAL science authors. We hope you will find something useful for addressing your own needs within these three different approaches.

17.1 Editor expectations of language use

Communicating meaning clearly is the crucial factor in scientific writing. Journal editors have often discussed the problem of clarity of communication in manuscripts. Some journals do use a copy-editor, but the general practice is that it is the author's responsibility to submit manuscripts in clearly understandable English. Many journal websites now make it clear this is the case. The quality of the science is the primary concern for journal editors; however, the science needs to be clearly understandable to the editor, the reviewers, and the ultimate readers.

It is in the best interests of authors to spend time on this aspect, because even errors that affect meaning to only a small extent can annoy readers and lead to an (unwarranted) impression that the science is inaccurate because the English is inaccurate. Many major publishers suggest that EAL authors may wish to use an online editing service pre-submission to help reach their language goals, and list some contacts for such fee-for-service editing on their websites. Even if you plan to use one of these services, or a face-to-face author's editor closer to home, it

Writing Scientific Research Articles: Strategy and Steps, Third Edition.
Margaret Cargill and Patrick O'Connor.
© 2021 John Wiley & Sons Ltd. Published 2021 by John Wiley & Sons Ltd.

remains important to ensure that you do as much language work as you can beforehand to ensure your meaning is as clear and unambiguous as possible and the English follows generally accepted usage in your discipline. This language work will likely save you money, as editors usually charge according to the time it takes to edit your manuscript. It will also help you avoid the risk that an editor may change your meaning inappropriately while "fixing" the English. Nevertheless, always take care to check the edited version of your manuscript before submission, to ensure the meaning is what you intended.

We have two suggestions for your writing in this context:

1 write short sentences first (two clauses only) and join them later if needed; and
2 aim to develop a repertoire of ways of expressing meanings that are useful in your discipline (a repertoire is a range of possibilities to choose from).

The following sections provide some ways to develop your repertoire.

17.2 Strategic (and acceptable!) language re-use: sentence templates

Recent research on EAL authors writing for publication in English has found that re-using language from other papers in the same field is a common strategy, but there is considerable discussion about when it is acceptable to do so and when the practice crosses into what can be called "textual plagiarism" (Flowerdew & Li 2007). What seems clear is that for science writing there is a divide in the way people think about the content – the science – and the way they think about the language used to express that content. The originality of the work is seen mostly to reside in the content: the data and their analysis and interpretation. This situation differs somewhat from that pertaining to writing in the humanities and social sciences, where the language is seen to form the argument, and therefore the content of the writing. Nevertheless, the very clear convention in academic writing in English is that to avoid the suspicion of plagiarism, authors should use their own words to paraphrase the findings or conclusions of other researchers, as well as citing the source of the information. Here we suggest a way in which EAL and other authors can be more confident about avoiding inappropriate language re-use while still taking advantage of the effective writing of other authors to develop their own repertoires. This option involves the construction of sentence templates for later re-use. We do this by separating the structure or framework of a sentence from the so-called "content chunks," the noun phrases.

To understand this concept, first read the following purpose statement, from an article by Li et al. (2000) entitled "Water use patterns and agronomic performance for some cropping systems with and without fallow crops in a semi-arid environment of northwest China," reprinted with permission from Elsevier:

As part of a long-term research effort aimed at establishing a sustainable rainfed farming system in the semi-arid and sub-humid regions of northwest China, this paper presents a detailed study on the water use patterns and agronomic performance for some cropping systems with and without fallow crops in a semi-arid environment. The objectives of this study were to: (1) determine the grain and aboveground biomass production and water-use efficiency of individual crops grown in the rotation; (2) analyze the seasonal and inter-annual patterns of soil water storage and utilization as well as water stress for the four major rotation crops such as winter wheat, corn, potato and millet; (3) determine the grain and aboveground biomass production and water-use efficiency for different rotation systems and evaluate the capacities of the rotation systems with and without fallow crops to utilize soil water storage in conjunction with seasonal precipitation; (4) establish whether the introduction of fallow crops into the wheat monoculture significantly influences the quantity of water stored in the soil that will be used by the subsequent wheat crop; and (5) discuss the characteristics of soil conservation for different rotation systems.

Source: Li, F., Zhao, S., & Geballe, G.T. (2000) Water use patterns and agronomic performance for some cropping systems with and without fallow crops in a semi-arid environment of northwest China. *Agriculture, Ecosystems and Environment, 79,* 129–142. © 2000, Elsevier

If we cross out all the noun phrases that relate just to this particular study, what remains is a series of frameworks that we call *sentence templates*:

As part of a long-term research effort aimed at ~~establishing a sustainable rainfed farming system in the semi-arid and sub-humid regions of northwest China~~, this paper presents a detailed study on ~~the water use patterns and agronomic performance for some cropping systems with and without fallow crops in a semi-arid environment~~. The objectives of this study were to: (1) determine ~~the grain and aboveground biomass production and water-use efficiency of individual crops grown in the rotation~~; (2) analyze ~~the seasonal and inter-annual patterns of soil water storage and utilization as well as water stress for the four major rotation crops of winter wheat, corn, potato and millet~~; (3) determine ~~the grain and aboveground biomass production and water-use efficiency for different rotation systems~~ and evaluate ~~the capacities of the rotation systems with and without fallow crops to utilize soil water storage in conjunction with seasonal precipitation~~; (4) establish whether ~~the introduction of fallow crops into the wheat monoculture significantly influences the quantity of water stored in the soil that will be used by the subsequent wheat crop~~; and (5) discuss ~~the characteristics of soil conservation for different rotation systems~~.

The frameworks or templates would look like this (NP = noun phrase):

As part of a long-term research effort aimed at [NP1], this paper presents [NP2]. The objectives of this study were to: (1) determine [NP3]; (2) analyze [NP4]; (3) determine [NP5] and evaluate [NP6]; (4) establish whether [NP7] significantly influences [NP8]; and (5) discuss [NP9].

N.B. You would only use this template, or any other template, if it enabled you to express the meanings you were trying to make. To help you decide what sorts of meanings they might be, it is useful to list and characterise the noun phrases that you crossed out to make a template, as demonstrated in Table 17.1.

We suggest that you continue to identify relevant sentence templates for yourself whenever you read a research paper for your work, in order to add to your repertoire. Take an extra 10 minutes or so after you have read a paper for its content to identify any useful sentence templates and record them in a special file or notebook. It may be useful to organise these notes according to the section of the paper where each template would be useful. Do Task 17.1 now, which focuses on Introductions.

Table 17.1 Relevant characteristics of noun phrases (NPs) for use in sentence templates.

Noun phrase	Characteristics
1 establishing a sustainable rainfed farming system in the semi-arid and sub-humid regions of northwest China	verb + ing + NP + in + [NP of location]
2 a detailed study on the water use patterns and agronomic performance for some cropping systems with and without fallow crops in a semi-arid environment	*a study* + on + NP + in + [NP of location]
3 the grain and aboveground biomass production and water-use efficiency of individual crops grown in the rotation	NP + of + [NP referring to features of study already introduced]
4 the seasonal and inter-annual patterns of soil water storage and utilization as well as water stress for the four major rotation crops of winter wheat, corn, potato and millet	NP + for + NP stating subjects of study
5 the grain and aboveground biomass production and water-use efficiency for different rotation systems	NP + for + NP stating subjects of study
6 the capacities of the rotation systems with and without fallow crops to utilize soil water storage in conjunction with seasonal precipitation	*the capacities of* [NP] to + verb + object
7 the introduction of fallow crops into the wheat monoculture	the introduction of + NP + into + NP
8 the quantity of water stored in the soil that will be used by the subsequent wheat crop	NP of measurement
9 the characteristics of soil conservation for different rotation systems	NP referring to types of conclusions expected from the study

---

Task 17.1   Drafting a sentence template for Stage 4 of an Introduction

1 Find the Introduction paragraph that contains Stage 4 in each of the Provided Example Articles (PEAs). To refresh your memory, Stage 4 is made up of the very specific sentences that present the purpose/objectives of the writer's study or outline its main activity or findings. What would the sentence templates look like? Draft them on a separate sheet of paper. Check your answer in the Answer pages.

2 Find Stage 4 in your Selected Article (SA). If it is suitable as the basis of a sentence template, construct one from it. Look at the noun phrases in your SA purpose statement. List them and note down any characteristics that will help you if you want to re-use the template in the future.

Discipline-specific noun phrases make up a very important part of the writing you need to do on your research. Identifying and learning them accurately is a very useful strategy for improving your writing skills. Here we present some grammatical details about noun phrases, and highlight one area of common difficulty associated with them.

A *noun phrase* is a group of words that does not include a finite verb (i.e. does not include a verb with a subject), built up around a single headword, which determines the grammatical relationships of the phrase to other elements of the sentence. The headwords are italicised in the following examples:

- the *mechanisms* of salt marsh succession;
- *interactions* involving carbohydrates;
- the seasonal and inter-annual *patterns* of soil water storage and utilization.

Note that long noun phrases can be made up of several smaller noun phrases, often joined together with prepositions.

A special case: noun–noun phrases

This kind of noun phrase can cause problems for EAL writers, in our experience. An example of a *noun–noun phrase* is "resource availability," which means "availability of resources." To shorten phrases like this, it is very common in scientific English for the second part (*of resources*) to be moved in front of the headword (*availability*). When this happens, the part that moves is generally written in its *singular* form (*resource*) and the preposition is omitted. (It is rare to find a possessive form with an apostrophe in such cases in science writing.) Similarly, "carbohydrate interactions" means "interactions involving carbohydrates." Table 17.2 contains some more examples, taken from the PEAs.

A good way to remember this construction is the following example:

food for *dogs* is *dog* food

Table 17.2 Examples of noun–noun phrases from the PEAs.

Noun–noun phrase	Extended form of the phrase
propagule pressure	pressure exerted by propagules
invasion success	success of invasions
field work	work conducted in the field
urchin disturbances	disturbances caused by urchins
legume root nodules	nodules on the roots of legumes
bacteroid activity	activity by bacteroids
bacteroid iron acquisition	acquisition of iron by bacteroids
soybean homologue	homologue in soybeans
lava flow hazard	hazard from flows of lava
A hazard map	a map of hazards
discharge rate estimates	estimates of rates of discharge

Task 17.2   Unpacking noun–noun phrases

Write down three noun–noun phrases commonly used in your research field. Next to each, unpack the phrase to explain what it actually means. For example:

> crop traits = traits exhibited by crops

Note the difference in the usage of singular and plural word forms in the two forms of the phrases. We suggest that you make a list of the noun–noun phrases you see used repeatedly in articles in your field, and learn them accurately, including whether the forms are singular or plural. This will help improve the accuracy of your writing considerably.

Using the noun phrase concept to read about unfamiliar areas of science

To summarise Sections 17.3 and 17.4, science writing is largely made up of sentence structures (templates), which are usable for many different areas of science, plus noun phrases, which are often specific to particular areas. Once you understand this concept, you will probably find it easier to read articles from areas of science with which you are not completely familiar. This is because you can skip over the unfamiliar noun phrases on your first reading, just concentrating on the sentence structures and main meanings. Then you can identify which noun phrases recur frequently, and use a dictionary or website to find out their meanings, if you need to know them. This will depend on your reason for reading the article. If you need to understand more about the area of research and are new to it, then you will probably need to look up many noun phrases. If you are reading the article only to find one specific piece of information, perhaps about the use of a method, you will need to look up fewer noun phrases. As you make your decisions about which ones to look up, remember to identify the headword of each noun phrase first, as this is the most important part for the sentence meaning.

The noun phrase idea can also help you to complete exercises in this book that involve writing about areas of science that are unfamiliar to you. For example, for readers who are unfamiliar with molecular biology and plant physiology, the PEA by Kaiser et al. (2003) may be challenging to read. Skipping over the complex noun phrases and focusing on the sentence structures will enable you to more easily do the exercises and understand the main point we are trying to teach. Of course, the same is true for the PEA by Britton-Simmons & Abbot (2008) for readers who are unfamiliar with marine biology studies, and for that by Ganci et al. (2012) for those unfamiliar with computer modelling and volcano studies.

17.4 Concordancing: a tool for developing your discipline-specific English

All languages contain words and phrases that are commonly associated with other words or phrases (e.g. theory and practice; genetically modified organisms; the effect of something on something else). These collocations (words that are commonly used together) can be identified and studied. If you want to identify and learn common collocations that are used in writing about your own research field, you need to study texts (examples of writing) specific to that field. In this section we introduce a type of software program that can help you do this in a systematic way: a concordancer.

What does a concordancer do?

A concordancer searches a group of texts (called a *corpus*) for all examples of a particular search item. It displays the results as lines of text across the screen, with the search term highlighted in the middle. Results can be sorted according to what is on the left or right of the search term (and one, two, or more words away from it), and this can provide data for your language learning. If the corpus you search is specific to your research field, you can search it in this way to improve your use of discipline-specific English.

In this section, we first provide an example of what can be learned from a concordancer search of a discipline-specific corpus (Task 17.3), and then explain how you can download freeware concordancing programs from the Internet and construct your own discipline-specific corpus.

Using concordancing software

For the language-development purposes we discuss here, readily available concordancing software programs from the Internet are ideal. A well-supported freeware example we recommend is AntConc (Anthony 2019), laurenceanthony.net/software/antconc/

Task 17.3   Getting familiar with concordancing

Look at the following concordancing search results, obtained by searching for the term "soil" in a corpus of articles from the field of soil science. Then read the questions and answers that follow.

> to utilise existing available *soil* water, unlike the perennial gr
> es (4 g oven dry wt basis) of *soil* were weighed into 40 ml polypr
> required 9 kg P/ha, whereas a *soil* with a high P sorption capacity
> concentration by 1 mg/kg on a *soil* with a low P sorption capacity
> 00, it was expected that this *soil* would have consistently been t
> capacity (PBC), which is the *soil*'s capacity to moderate changes
> and buffering capacity of the *soil*-an attempt to test Schofield's
> nisms that are present in the *soil*-plant microcosm environment. T
> etermined in a growth-chamber *soil*-plant microcosm study. Nodding
> 84) Lime and phosphate in the *soil*-plant system. Advances in Agro
> a where crops rely heavily on *soil*-stored water accrued in summer
> fertility on these particular *soil*s. Although this aberration has
> over in a range of allophanic *soil*s amended with 14 Clabelled gluc
> alues for 9 different pasture *soil*s, 6 and 12 months after P fert

Q1 Is soil countable, uncountable, or both in these examples?

A1 Both. Countable examples include "a soil with a high P sorption capacity" and "9 different pasture soils"; an uncountable usage can be seen in "samples of soil were weighed."

Some of these usages are different from those found in everyday English, where soil is always uncountable. From this example, you can get an idea of how a concordancer search of a discipline-specific text collection can help you identify English usages that are specific to that discipline.

Q2 How many different ways is the word "soil" used in these examples?

A2 Quite a few! For example, as well as its usage as a countable and uncountable noun, it is used in noun–noun phrases, both as the headword ("pasture soils") and as the adjective-equivalent ("soil water"), and in hyphenated adjectival constructions ("soil-stored water") and noun–noun phrases ("soil–plant microcosm").

described by its creator as a "freeware corpus analysis toolkit for concordancing and text analysis." Helpful training videos and downloadable guides are also available from the same site.

Once you have a concordancing program installed, you will need a corpus – a collection of suitable text to analyse. Our suggestion is that you construct a corpus of English-language journal articles from your own discipline(s), so you can search it for the use of words or phrases you need in your scientific writing. This will provide data, on your own desktop, for your ongoing learning of the specific English phrases and expressions used in your discipline.

Making a corpus

To be most useful, a corpus needs to consist of documents from your own subdiscipline, of the type you are aiming to write. A useful corpus for EAL scientists wanting to write articles for international publication would be 20–30 published research articles in their particular field. Our suggestion is that the articles used for a corpus be selected or approved by supervisors or leaders of research groups, to ensure that they

- are from reputable journals in the field;
- are well-written, by authors using English as a first language or at a comparable level;
- cover a suitable range of subtopics within the field, to give a good range of language usage; and
- cover the required range of types of writing (e.g. including or excluding review articles, as desired).

Preparing documents for a corpus

To be searchable by a concordancing program, the texts must be saved in text-only format (.txt). If you can download the articles from their source journal or database in html format (full text, not pdf), saving the files as text-only (.txt) is a straightforward operation. Alternatively, if the selected articles are available in Microsoft Word format (e.g. if the author is willing to provide the text in this format), then the same process is possible. In both cases, delete the tables and figures, the author biodata, and the reference lists before saving as .txt files. If the articles are in .pdf format, then a somewhat tedious set of steps needs to be followed (see later). All files should be placed in a single folder (directory) on your computer for ease of searching.

Copyright issues

Making a single electronic copy for use with a concordancer is comparable to making a single copy for research use.

Preparing text in pdf format for use with a concordancer

Two approaches are possible for this task. The first is to use a conversion program such as AntFileConverter (Anthony 2017). A disadvantage of this method is that page headers and footers are also converted, so the .txt file contains text that is not part of the original article. To make a clean file, these headers and footers need to be removed manually.

In the second method, a copy/paste procedure must be followed to convert the .pdf text to a text-only format. Only the written part of the article is needed, so do not copy biodata, tables and figures, reference lists, or acknowledgements, and do not include headers and footers.

- Open the file in Adobe Acrobat Reader.
- Select as much text as you can without including unwanted items such as headers and footers, page numbers, tables/figures, or the reference list.
- Copy the text (Ctrl + C).
- Open your word processor (e.g. Microsoft Word).
- Paste the text into a new document (Ctrl + V).
- Repeat the steps of selecting, copying, and pasting until all the relevant parts of the paper are copied.
- Save the file as text only (.txt) and name the file before saving.
- Edit the text file as necessary (remove unwanted material).

Complete Task 17.4 to further your understanding of this tool.

Task 17.4   Practice with concordancing

Practice using the concordancer (or read carefully) to examine the texts in your corpus of journal articles in order to answer the following questions:

1 Do article authors begin sentences with "Also"?
2 What about "In addition"?
3 How else is "addition" used?
4 Do authors use "I" or "we"? If so, in which sections of the articles (e.g. Methods, Discussion)?
5 What constructions are used with the verb "affect"?
6 What verbs are used with the noun "role"? And what prepositions are used after this word?

Now, think of other searches that you could try to find answers to language questions that have arisen for you as you write about your research. Discuss with a colleague or teacher, if possible.

17.5 Using the English articles (a/an, the) appropriately in science writing

For many of you who use EAL, the problem of using articles appropriately has been a constant since your early days of English learning. You may have seen the rules explained in many different ways, and learned them over and over again. Why then have we chosen to discuss this issue again here? Precisely because it is so difficult to master, especially for EAL users whose first language does not contain articles, and because it is often highlighted by journal editors and referees as needing attention in submitted manuscripts.

Indeed, in our experience editors and referees who speak English as a first language (EL1), and who therefore learned article usage by immersion "at their mother's knee," may have limited understanding of the complexity of this part of the English language system. This complexity is reflected in the fact that effective computer software to identify or correct article errors has not yet, to our knowledge, been developed. This lack reflects the degree to which the use of English articles with any noun phrase depends on the meaning of the phrase in its particular context in the sentence, especially whether the noun phrase is used there in a generic sense or a specific sense. This question (generic or specific) relates also to the problems of meaning that can occur when articles are used inappropriately. It is therefore with the generic/specific question that we begin our discussion of article use.

Generic noun phrases

Generic noun phrases refer to any – or all – members of a particular class or category of living things, objects, or concepts. There are four ways to write these generic noun phrases in English:

1 If the noun is *countable*, you can make it generic by writing it in its plural form and not using any article.
2 An alternative when the noun is *countable* is to make it generic by using its singular form with the article *a* or *an*.

   e.g. *Healthy crops* can contribute substantial cadmium to *human diets*.
   A healthy crop can contribute substantial cadmium to human diets.

3 When the noun you want to use is *uncountable*, you make it generic by omitting any article. (Remember: uncountable nouns never have a plural form.)

   e.g. *Cadmium* exists in soils in many forms.
   Manipulation of *soil pH* can be effective in managing *Cd contamination*.

4 English has another possible way of making generic noun phrases which you need to recognize. Sometimes, a singular countable noun carries the generic meaning when used with the definite article *the*. This is often used when referring to living things or common machinery or equipment. (It is usually also possible to substitute the plural form of the word without an article – remembering also to change the verb to agree, of course.)

   e.g. *The earthworm* can be found in many types of soil. (or *Earthworms can. . .*)
   The computer has become an important tool for researchers. (or *Computers have. . .*)

N.B. For science writing in particular, it is important to remember that as long as you continue talking about a noun as a concept or general class (any or all of them), the noun *stays generic* (i.e. you may have to unlearn the general rule that says a noun is specific after it has been used once in a passage of writing). Complete Task 17.5 now.

Task 17.5   Generic noun phrases

In the first paragraph of the Introduction to the PEA by Kaiser et al. (2003), reproduced here, underline examples of generic noun phrases using both countable and uncountable nouns.

> Legumes form symbiotic associations with N_2-fixing soil-borne bacteria of the *Rhizobium* family. The symbiosis begins when compatible bacteria invade legume root hairs, signalling the division of inner cortical root cells and the formation of a nodule. Invading bacteria migrate to the developing nodule by way of an "infection thread," comprised of an invaginated cell wall. In the inner cortex, bacteria are released into the cell cytosol, enveloped in a modified plasma membrane (the peribacteroid membrane (PBM)), to form an organelle-like structure called the symbiosome, which consists of bacteroid, PBM and the intervening peribacteroid space (PBS; Whitehead and Day, 1997). The bacteria, subsequently, differentiate into the N_2-fixing bacteroid form. The symbiosis allows the access of legumes to atmospheric N4, which is reduced to NH_2^+ by the bacteroid enzyme nitrogenase. In exchange for reduced N, the plant provides carbon to the nodules to support bacterial respiration, a low-oxygen environment in the nodule suitable for bacteroid nitrogenase activity, and all the essential nutritional elements necessary for bacteroid activity. Consequently, nutrient transport across the PBM is an important control mechanism in the promotion and regulation of the symbiosis. (Source: Kaiser, B.N., Moreau, S., Castelli, J., et al. The soybean NRAMP homologue, GmDMT1, is a symbiotic divalent metal transporter capable of ferrous iron transport. *The Plant Journal*, 35, 295–304. © 2003, John Wiley & Sons.)

Check your answers in the Answer pages.

Specific noun phrases

Specific noun phrases refer to particular, individual members of a class or category, rather than the class as a whole. The reader and the writer both know already which one or ones are being referred to. This requires the use of *specific noun phrases*, which involve the definite article *the*. There are three different reasons why a specific noun phrase may be required:

1   The noun phrase is specific because the phrase is referring to shared or assumed knowledge of one particular referent (= the thing being referred to).

    e.g. In recent years the growth of desert areas has been accelerating in *the world*.

2   The noun phrase is specific because the phrase is pointing back to old information already introduced to the reader.

    e.g. A pot experiment was conducted in an acid soil. *The experiment* showed. . .

3   The noun phrase is specific because the phrase is pointing forward to information that specifies which one or ones are being referred to.

    e.g. The aim of this study was to investigate *the effect* of liming on Cd uptake.

N.B. It is worth noting that when the structure NP1 + of + NP2 is used, the first noun phrase will be specific (i.e. have *the* in front of it) about 85% of the time. It is therefore a good idea to always use *the* in this situation, unless you are very sure that the extended noun phrase (the two noun phrases joined with *of*) is generic for some reason. Do Task 17.6 now.

Task 17.6   Specific noun phrases

Reread the Introduction paragraph from the PEA by Kaiser et al. (2003) and draw a square around each specific noun phrase. Discuss with a colleague why each one is specific.

Check your answers in the Answer pages.

Summary flow chart for deciding on article use

Many EAL writers find the flow chart presented in Figure 17.1 helpful when they have to decide which form of the article to use with a noun phrase in a particular sentence. Do Task 17.7 now.

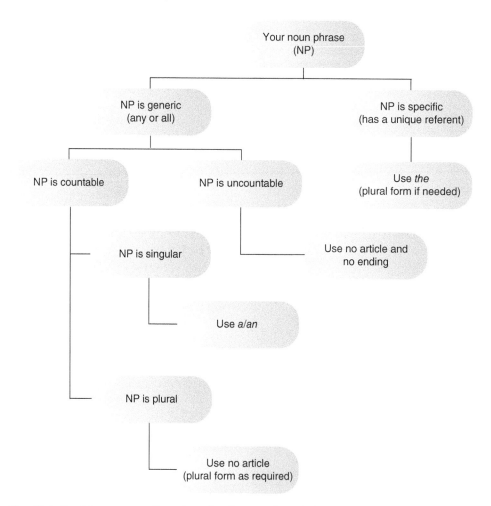

Fig. 17.1 Decision-support flow chart for the use of English articles (*a/an/the*). *Source*: Based on Weissberg, R. & Buker, S., *Writing Up Research: Experimental Research Report Writing for Students of English*. Prentice Hall Regents.

Task 17.7   Articles and plurals in a science paragraph

Consulting the flow chart in Figure 17.1, fill in each blank space in the following extract from the PEA by Britton-Simmons & Abbott (2008) with the plural marker -s, a, an, or the. Some of the blanks do not require filling in.

_____ propagule pressure is widely recognized as _____ important factor that influences_____ invasion success._____ previous studies suggest that probability of _____ successful invasion increases with _____ number of propagules released, with _____ number of introduction attempts, with _____ introduction rate, and with _____ proximity to _____ existing populations of invaders. Moreover, _____ propagule pressure may influence _____ invasion dynamics after _____ establishment by affecting _____ capacity of _____ non-native species to adapt to their new environment. Despite its acknowledged importance, _____ propagule pressure has rarely been manipulated experimentally and _____ interaction of _____ propagule pressure with _____ other processes that regulate _____ invasion success is not well understood.

Check your answers in the Answer pages.

17.6 Using "which" and "that"

Problems with the relative pronouns *which* and *that*, and whether to use commas with them, are a common error we see in editing scientists' writing. The explanation here is designed to help you understand and remember how these two words are used.

Example 1: Land *which is surrounded by water* is an island.

The italicised portion of Example 1 is the relative clause. In this example, the relative clause is essential to the meaning of the sentence, because if it were omitted the sentence would read "Land is an island." This does not make sense, because only land which is surrounded by water is called an island. Thus the relative clause defines which land the sentence refers to: it is a defining (or restrictive) relative clause.

The important points to note about defining relative clauses are as follows:

- Practice differs in terms of the relative pronoun that can begin a defining clause. Some writers (and teachers of writing) feel comfortable with either *which* or *that*. Others permit only *that* in this clause type. However, the publisher of your accepted article may have in-house rules on this topic.
- Defining clauses have no commas separating them from the rest of the sentence in which they are placed.

Example 2: Tasmania, *which is surrounded by the waters of Bass Strait*, is an island of great natural beauty.

In this example, the relative clause is not essential to the basic meaning of the sentence. If it were omitted, the sentence would read "Tasmania is an island of great natural beauty," and this makes sense. The relative clause is adding extra, non-essential information and is thus a non-defining (non-restrictive) relative clause. Another way to work out if a relative clause is non-defining is to try inserting the phrase "by the way" after the *which*. If this addition sounds acceptable, the clause is non-defining.

The points to note about non-defining relative clauses are these:

- They can only begin with *which*.
- They are separated from the rest of their sentence by commas: two commas if they occur in the middle of the sentence as in our example, or one comma if they come at the end of a sentence.

N.B. The same distinction regarding punctuation holds when the *which/that* + *verb* is omitted, forming a phrase.

> **Examples:** Tasmania, surrounded by the waters of Bass Strait, is an island of great natural beauty.
>
> Land surrounded by water is an island.

Task 17.8 provides some practice in punctuating sentences of the types we have been discussing here.

Task 17.8   Punctuation with "which" and "that"

Punctuate the following examples:

1 Lime which raises the pH of the soil to a level more suitable for crops is injected into the soil using a pneumatic injector.
2 Manipulation which involves adding or deleting genetic information is referred to as genetic engineering.
3 Non-cereal phases which are essential for the improvement of soil fertility break disease cycles and replace important soil nutrients.
4 Senescence which is the aging of plant parts is caused by ethylene that the plant produces.
5 Opportunities that arise from the economically buoyant nature of domestic wine production must be identified and carefully assessed.
6 Seasonal cracking which is a notable feature of this soil type provides pathways at least 6 mm wide and 30 cm deep that assist in water movement into the subsoil.
7 Plants which experience waterlogging early in their development would be expected to have a much shallower root system than non-waterlogged plants.
8 Yellow lupin which may tolerate waterlogging better than the narrow-leafed variety has the potential to improve yields in this area.
9 Lucerne is a drought-hardy perennial legume which produces high-quality forage.

Check your answers in the Answer pages.

CHAPTER 18

Writing funding proposals

Obtaining funding for further work is a vital part of most research careers, and in many cases this funding is awarded on the basis of written grant applications that outline proposals designed to meet the criteria and priorities of the body providing the funds. The process of deciding which proposals will be funded is highly competitive, as there are often many more strong proposals than can be funded. In this regard, the situation of a proposal submitter is similar to that of an author submitting a paper to a high-ranking journal. Thus, we focus in this chapter on applying to the task of writing a funding proposal the principles of strategy development and effective writing that have been discussed elsewhere in the book. We recognise that this approach means we will not cover all points that are relevant to the assessment of funding proposals, in particular the strength of the scientific work proposed. It is also important to note that genre analysis results of the sort that underpin the other parts of the book are less widely available in the case of funding proposals, perhaps due to difficulties in accessing suitable collections of proposals for analysis or the great diversity of funding proposal requirements – but see Flowerdew (2016) for a comparison of findings on types of information included in proposals. The discussion here is based rather on the application of appropriate lessons from genre analysis for related sections of scientific articles, and on the experience of the authors and of the "critical readers" in various scientific fields who have provided feedback on drafts of the chapter. Overall, applying the suggestions provided here can potentially both strengthen proposals and streamline their production.

Timing issues are important, as the production of a strong proposal is extremely time-consuming. A first question to be answered is this: Is now the time for you to write this application? Important considerations here include whether it would be better to finish writing the paper you have underway first, and whether you have left enough time to develop a quality application. It may also be necessary to consider whether you have the necessary results from preliminary or prior research work to justify the application at this time. To help address the time-pressure issue, a useful strategy is to keep a list of projects on file with enough detail fleshed out to allow a running start when grant applications open. The steps discussed in this chapter may provide pointers for the types of detail to include in your own case.

A wide range of grant application processes and selection criteria are used by different funding bodies, and it is thus important to test the applicability of the suggestions in this chapter to your own situation. Types of grants and application formats, criteria, and processes differ according to scientific subdiscipline, and cultural/national differences can also be important. However, there are two overarching imperatives for any grant proposal: (1) the proposal is for important

Writing Scientific Research Articles: Strategy and Steps, Third Edition.
Margaret Cargill and Patrick O'Connor.
© 2021 John Wiley & Sons Ltd. Published 2021 by John Wiley & Sons Ltd.

work which is scientifically sound and appropriate and which it is feasible to undertake at this time; and (2) the application guidelines are followed and all important information is supplied, while information not required is omitted. Time taken researching the specific requirements of funding bodies for funding programmes in your own field is time well spent.

18.1 A process for preparing and submitting a funding proposal

Five strategic planning questions for project proposals

Echoing the four questions for manuscript planning presented in Chapter 4, the following five questions will assist in the initial planning of a funding proposal. Answering them in writing will provide information to help you target an appropriate funding source and plan an effective application.

1. What will this particular project aim to achieve? (specific, measurable outcomes)
2. Why are these outcomes important now?
3. Who are the members of the project team that can deliver these outcomes to a high standard within the required timeframe?
4. Which funders will potentially be interested in supporting this research project?
5. Why should they fund this project in preference to others that may be submitted?

Matching your project to the most likely funding body

A number of issues need to be taken into consideration here. Examples include

- the type of research planned (e.g. is it focused on a theoretical or applied contribution?);
- the problem to be addressed;
- the level of benefit/significance envisaged, including the chance of "win–win" research (e.g. where results supporting the null hypothesis still represent a contribution);
- the duration of funding support available, in relation to the size of the project; and
- the value ascribed by the funder to collaboration (and with whom) and to a team's track record.

This matching process requires careful analysis of all documentation provided by the funding body, including study of the titles and aims or summaries of projects funded in the past, if these are available. Check also for any extra priority areas or emphases that may be apparent from the promotional material or annual reports for the grant, in terms of where outcomes are published or publicised and what aspects of projects are featured most strongly in the reporting. If your project can be described in comparable terms, then you have a useful "hook" for use in writing a proposal to this funding body.

At this point, once the target funder is decided, it is useful to sequence the necessary stages for development of the application and record timing issues, in order to improve efficiency for both the present and future funding opportunities. Stages include review of grant criteria and requirements; institutional requirements and

Deciding on the take-home message

A basic format for this proposal segment is the title plus a one-paragraph summary. We suggest that you write the project aims first, which will form the key content towards the end of the summary paragraph. Then add introductory sentences providing only essential background information, to connect the aims to the funder's priorities and to show the gap to be addressed: why this project needs to be done now. This summary can use the same argumentation stages that are highlighted in Chapter 8 for research article Introductions, with the emphasis on Stage 4 (aims, written first), Stage 3 (the gap, research niche, remaining problem, or need for extension), and Stage 1 (claims of importance in the context of the funder's priorities). Later, when all other sections are complete, review the aims section and ensure that the research gap, methods, and any preliminary results align tightly with it. The title can be developed at the end of the process using the approach recommended in Chapter 10 for journal articles.

Next, write bullet points only for each main section of the proposal format used by the target funder. Use these points to highlight aspects you think will be most relevant in meeting the funder's criteria. Now you are ready for a first round of peer feedback.

Feedback from mentors/colleagues – Round 1

Seek advice on your proposal "skeleton" from experienced researchers with expertise relevant to the proposal topic – especially any who have received funding in the past from the organisation you are targeting. Ask them about your match with funder priorities, as well as things to focus on in the development of a full proposal.

Once the feedback is in, decide if the issues raised can be addressed satisfactorily with a reasonable expenditure of time and effort. If so, plan your order of attack, input from team members, and timelines to allow formal internal review and feedback before the submission deadline. Refer to your stage sequence, and refine it as appropriate.

Assessment criteria analysis

With your team, analyse the assessment criteria provided by the funding body, including the percentage weightings for different criteria if these are available. Decide in which section(s) you can provide evidence that your proposal meets each criterion, and what evidence is required. Gather the evidence not already to hand and ensure it is incorporated in the relevant sections.

Compilation and revision of sections: the first complete draft

Combine the section drafts prepared by team members; check that each follows closely any guidelines or instructions provided by the funding body and incorporates the evidence you have gathered relating to assessment criteria.

Prepare an honest list of "problems and pitfalls" in your proposed research. If you have thought of it, chances are the grant reviewers will think of it too. Make sure you have demonstrated in the proposal how these problems and pitfalls will be overcome or managed.

As you revise the sections within the project team, pay particular attention to the strategies for creating logical flow in science writing (Chapter 8). These are particularly important for grant assessors (your target reading audience), because they may not be experts in the specifics of your research topic and are thus likely to miss connections between points unless they are made clear in the way the points are put together. This is notably the case for first-round assessors in some big national grant schemes. Make sure also that you adhere to any page or word limits.

Formal internal review – Round 2

Seek formal internal review of your draft in good time, especially from scientists who have been funded by or worked as an assessor for the funding body you are applying to. Formal internal review aims to simulate the selection process your application will go through when it reaches the funding body. You want reviewers in your institution to find any mistakes or weaknesses in the application so that you can deal with them before you submit it. Internal review may not be a requirement of your institution, but you can ask colleagues to review your application as though they were acting on behalf of the funding body. Provide reviewers with a copy of the criteria if they are available. If the specific criteria are not available, then two general questions are these:

- Would you give this team the money it has requested?
- If not, what changes could be made to the application to strengthen its chances?

Seek separate feedback on the language of your draft; ask the reviewers to comment on how easy it is to follow and whether they had to reread any parts to understand the message clearly. Decide on the most appropriate way to respond to all feedback you receive before submitting your application. However, remember that not all institutional reviewers are experts in your field and they may misunderstand or misinterpret what you have written. You need to decide which feedback is relevant and important.

Towards the next time

Whether or not you are successful with your application, make sure to record any useful feedback you receive from the funding body and use it to enhance the strategies you will employ for the next grant application you write. For each element of the feedback received, decide at which preparation stage it could be applied for best effect, and ensure that these decisions are addressed next time around. Key ongoing documents that should be kept up to date to assist in more efficient grant application writing include:

- the fleshed-out list of potential projects described at the start of this chapter; and
- the sequence of project development steps and timing issues for each described earlier in this section.

Make sure also that you take full advantage of any support offered by relevant sections of your institution for maintaining awareness of current funding opportunities and preparing competitive applications. A further useful reference is Chapter 10 in Johnson (2011), freely downloadable online, which lists international funding sources and provides tips for presenting yourself as a researcher with strong credibility.

18.2 Easy mistakes to make

The following list of potential mistakes is not comprehensive, but indicates areas where haste or lack of planning may undermine or disqualify your application:

- Ignoring a guideline about the type and amount of content to be supplied for any application criterion. Funding bodies may receive many applications, and they expect high-quality applications which demonstrate professionalism in following the instructions provided.
- Failing to understand the funders' interests, objectives, and priorities. It is their money; they will choose who to give it to.
- Failing to target the language in each section to the audience that will assess that section (e.g. an Introduction section may be read by a general or non-expert audience, while a Methods section may be read by a discipline or subdiscipline expert).
- Providing unrealistic timelines, scopes of work, or budgets. The funder wants a high likelihood of getting what they are promised within a specified time; they will probably be able to tell if promises are unrealistic.
- Concentrating too much on explanations of what will be done, rather than including enough information about why the work is important, and why now. Funding bodies want to invest in success.
- Failing to make explicit, within the text about your project, key answers to funders' questions/criteria. Reviewers do not want to hunt for critical information buried in your prose.
- Unhelpful use of jargon, buzzwords, acronyms, clichés, or wordy, distracting prose which does not aid clear, precise communication against the criteria.
- Submitting text that contains imprecise language, typographical or grammatical errors, or incorrect notation. Edit carefully, because sloppy proposals suggest sloppy researchers.

We suggest reading this list several times during the process of preparing and revising your funding proposal. The goal is to keep your focus firmly on the audience for the document – the funding body – and thereby increase your chances of meeting their expectations to a high degree.

Provided example articles

CHAPTER 19

PEA1: Kaiser et al. (2003)

Reproduced from Kaiser BN, Moreau S, Castelli J, et al. The soybean NRAMP homologue, GmDMT1, is a symbiotic divalent metal transporter capable of ferrous iron transport. *Plant J*. 2003;35(3):295–304. doi:10.1046/j.1365-313x.2003.01802.x.

Writing Scientific Research Articles: Strategy and Steps, Third Edition.
Margaret Cargill and Patrick O'Connor.
© 2021 John Wiley & Sons Ltd. Published 2021 by John Wiley & Sons Ltd.

The Plant Journal (2003) 35, 295-304

doi: 10.1046/j.1365-313X.2003.01802.x

The soybean NRAMP homologue, GmDMT1, is a symbiotic divalent metal transporter capable of ferrous iron transport

Brent N. Kaiser[1], Sophie Moreau[2], Joanne Castelli[3], Rowena Thomson[3], Annie Lambert[2], Stéphanie Bogliolo[4], Alain Puppo[2], and David A. Day[3,*]

[1]*School of Agricultural Sciences, Discipline of Wine & Horticulture, The University of Adelaide, Urrbrae, South Australia, Australia*

[2]*Laboratoire de Biologie Végétale et Microbiologie, CNRS FRE 2294, Université de Nice-Sophia Antipolis, Parc Valrose, 06108, Nice, cédex 2, France*

[3]*Biochemistry & Molecular Biology, School of Biomedical & Chemical Sciences, University of Western Australia, Crawley, WA, 6009, Australia, and*

[4]*Laboratoire de Physiologie des Membranes Cellulaires, UMR 6078 CNRS-Université de Nice-Sophia Antipolis, 284 chemin du Lazaret, 06230 Villefranche sur Mer, France*

Received 9 December 2002; revised 24 April 2003; accepted 7 May 2003.
*For correspondence (fax +61 08 9380 1148; e-mail dday@cyllene.uwa.edu.au).

Summary

Iron is an important nutrient in N_2-fixing legume root nodules. Iron supplied to the nodule is used by the plant for the synthesis of leghemoglobin, while in the bacteroid fraction, it is used as an essential cofactor for the bacterial N_2-fixing enzyme, nitrogenase, and iron-containing proteins of the electron transport chain. The supply of iron to the bacteroids requires inital transport across the plant-derived peribacteriod membrane ehich physically separates bacteroids from the infected plant cell cytosol. In this study, we have identified *Glycine max divalent metal transporter 1* (*GmDmt1*), a soybean homologue of the NRAMP/Dmt1 family of divalent metal ion transporters. *GmDmt1* shows enhanced expression in soybean root nodules and is most highly expressed at the onset of nitrogen fixation in developing nodules. Antibodies raised against a partial fragment of GmDmt 1 confirmed its presence on the peribacteriod membrane (PBM) of soybean root nodules. GmDmt 1 was able to both rescue growth and enhance ^{55}Fe(II) uptake in the ferrous iron transport deficient yeast strain (*fet3fet4*). The results indicate that GmDmt1 is a nodule-enhanced transporter capable of ferrous iron transport across the PBM of soybean root nodules. Its role in nodule iron homeostasis to support bacterial nitrogen fixation is discussed.

Keywords: iron, NRAMP, nitrogen fixation, soybean, symbiosome.

Introduction

Legumes form symbiotic associations with N_2-fixing soil-borne bacteria of the *Rhizobium* family. The symbiosis begins when compatible bacteria invade legume root hairs, signalling the division of inner cortical root cells and the formation of a nodule. Invading bacteria migrate to the developing nodule by way of an 'infection thread', comprised of an invaginated cell wall. In the inner cortex, bacteria are released into the cell cytosol, enveloped in a modified plasma membrane (the peribacteroid membrane (PBM)), to form an organelle-like structure called the symbiosome, which consists of bacteroid, PBM and the intervening peribacteroid space (PBS; Whitehead and Day, 1997). The bacteria, subsequently, differentiate into the

N_2-fixing bacteroid form. The symbiosis allows the access of legumes to atmospheric N_2, which is reduced to NH_4^+ by the bacteroid enzyme nitrogenase. In exchange for reduced N, the plant provides carbon to the nodules to support bacterial respiration, a low-oxygen environment in the nodule suitable for bacteroid nitrogenase activity, and all the essential nutritional elements necessary for bacteroid activity. Consequently, nutrient transport across the PBM is an important control mechanism in the promotion and regulation of the symbiosis.

Micronutrients such as iron are essential for bacteroid activity and nodule development. The demand for iron increases during symbiosis (Tang *et al.*, 1990), where the

© 2003 Blackwell Publishing Ltd

153

(a)

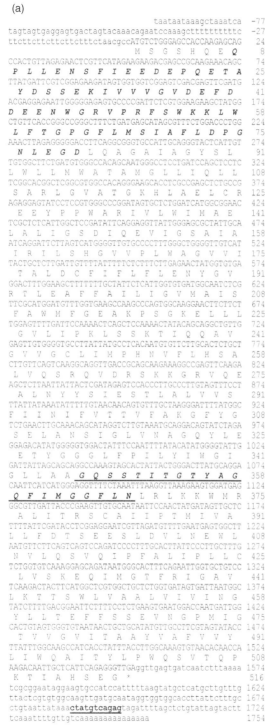

(b)

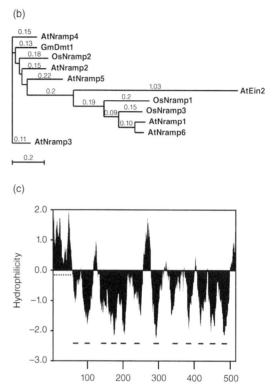

(c)

metal is utilised for the synthesis of various iron-containing proteins in both the plant and the bacteroids. In the plant fraction, iron is an important part of the heme moiety of leghemoglobin, which facilitates the diffusion of O_2 to the symbiosomes in the infected cell cytosol (Appleby, 1984). In bacteroids, there are many iron-containing proteins involved in N_2 fixation, including nitrogenase itself and cytochromes used in the bacteroid electron-transport chain. In the soil, iron is often poorly available to plants as it is usually in its oxidised form Fe(III), which is highly insoluble at neutral and basic pH. To compensate this, plants have developed two general strategies to gain access to iron from their localised environment. Strategy I involves secretion of phytosiderophores that aid in the solubilisation and uptake of Fe(III), while strategy II involves initial reduction of Fe(III) to Fe(II) by a plasma membrane Fe(III)-chelate reductase, followed by uptake of Fe(II) (Romheld, 1987). The mechanism(s) involved in bacteroid iron acquisition within the nodule have been investigated at the biochemical level, and three activities have been identified (Day *et al.,* 2001). Fe(III) is transported across the PBM complexed with organic acids such as citrate, and accumulates in the PBS (Levier *et al.,* 1996; Moreau *et al.,* 1995), where it becomes bound to siderophore-like compounds (Wittenberg *et al.,* 1996). Fe(III) chelate reductase activity has been measured on isolated PBM, and Fe(III) uptake into isolated symbiosomes is stimulated by Nicotinamide Adenine Dinucleotide (NADH), reduced form (Levier *et al.,* 1996). However, Fe(II) is also readily transported across the PBM and has been found to be the favoured form of iron taken up by bacteroids (Moreau *et al.,* 1998). The proteins involved in this transport have not yet been identified.

Two classes of putative Fe(II)-transport proteins (Irt/Zip and Dmt/Nramp) have been identified in plants (Belouchi *et al.,* 1997; Curie *et al.,* 2000; Eide *et al.,* 1996; Thomine *et al.,* 2000). The Irt/Zip family was first identified in *Arabidopsis* by functional complementation of the yeast Fe(II) transport mutant DEY1453 *(fet3fet4;* Eide *et al.,* 1996). *AtIrt1* expression is enhanced in roots when grown on low iron (Eide *et al.,* 1996), and appears to be the main avenue for iron acquisition in *Arabidopsis* (Vert *et al.,* 2002). Recently, a soybean Irt/Zip isologue, GmZip1, was identified and localised to the PBM in nodules (Moreau *et al.,*

2002). GmZip1 has been characterised as a symbiotic zinc transporter, which does not transport Fe(II). The second class of iron-transport proteins consists of the Dmt/Nramp family of membrane transporters, which were first identified in mammals as a putative defence mechanism utilised by macrophages against mycobacterium infection (Supek *et al.,* 1996; Vidal and Gros, 1994). Mutations in Nramp proteins in different organisms result in varied phenotypes including altered taste patterns in *Drosophila* (Rodrigues *et al.,* 1995), microcytic anaemia (mk) in mice and belgrade rats (Fleming *et al.,* 1997) and loss of ethylene sensitivity in plants (Alonso *et al.,* 1999). The rat and yeast NRAMP homologues (DCT1 and SMF1, respectively) have been expressed in *Xenopus* oocytes and shown to be broad-specificity metal ion transporters capable of Fe(II), amongst other divalent cations, transport (Chen *et al.,* 1999; Gunshin *et al.,* 1997). The plant homologue, AtNramp1, complements the growth defect of the yeast Fe(II) transport mutant DEY1453, while other *Arabidopsis* members do not (Curie *et al.,* 2000; Thomine *et al.,* 2000). Interestingly, AtNramp1 overexpression in *Arabidopsis* also confers tolerance to toxic concentrations of external Fe(II) (Curie *et al.,* 2000), suggesting, perhaps, that it is localised intracellularly.

In this study, we have identified a soybean homologue of the Nramp family of membrane proteins, GmDmt1;1. We show that GmDmt1;1 is a symbiotically enhanced plant protein, expressed in soybean nodules at the onset of nitrogen fixation, and is localised to the PBM. GmDmt1;1 is capable of Fe(II) transport when expressed in yeast. Together, the localisation and demonstrated activity of GmDmt1;1 in soybean nodules suggests that the protein is involved in Fe(II) transport and iron homeostasis in the nodule to support symbiotic N_2 fixation.

Results

Cloning of GmDmt1;1

A partial cDNA of GmDmt1;1 was identified from a 6-week-old soybean nodule cDNA library during a 5'-RACE PCR experiment designed to amplify the N-terminal sequence of a putative NH_4^+ transporter, GmAMT1. Subsequent PCR

Figure 1 Sequence analysis.
(a) Nucleotide and the deduced amino acid sequence of GmDmt1;1. Amino acids italicised and in bold represent the N-terminal region of GmDmt1;1 used for the generation of the anti-GmDmt1;1 antisera. Consensus Dmt transport motif (bold italic underlined amino acids) and putative iron-responsive element (IRE; bold underlined) are indicated.
(b) Phylogenetic tree of selected members of the Dmt/Nramp family found in plants AtNramp1 (AF165125), AtNramp2 (AF141204), AtNramp3 (AF202539), AtNramp4 (AF202540), AtNRAMP5 (CAC27822), AtNramp6 (CAC28123), AtEin2 (AAD41076), OsNramp1 (S62667), OsNramp2 (AAB61961), OsNramp3 (AAC49720). The phylogenetic tree was drawn using MacVector (Accelrys) after comparison of deduced amino acid sequences using the CLUSTAL W method. The phylogram was built using the neighbour-joining method and best-tree mode. Distances between proteins were estimated using the Poisson-correction algorithm.
(c) Hydropathy analysis of the deduced amino acid sequence of GmDmt1;1 calculated using the Kyte and Doolittle algorithm with an amino acid window size of 19. Putative transmembrane spanning regions are indicated with horizontal bars. Dashed bar indicates hydrophilic section of protein used to generate anti-GmDmt1 antisera.

experiments identified a full-length 1849-bp cDNA, which was cloned and sequenced (Figure 1a) (accession no. AY169405). Analysis of the GmDmt1;1 nucleotide sequence identified an open-reading frame of 516 amino acids encoding for a putative protein of approximately 57 kDa (Figure 1a). A BLAST search analysis of the GmDmt1;1 amino acid sequence identified significant homology (approximately 29% identity; approximately 46% similarity) to the amino acid sequences of six members of the *Arabidopsis* Nramp family (excluding AtEin2) of divalent metal ion transporters (Figure 1b). Hydropathy analysis (Kyte and Doolittle, 1982) of the encoded amino acids identified a protein with 12 putative transmembrane-spanning regions (Figure 1c). Between transmembrane segments 8 and 9, there is a conserved transport motif (5′-GQSSTITGTYAGQ-FIMGGFLN-3′), common among Nramp/Dmt homologues (Figure 1a). In the 3′-untranslated region of *GmDmt1;1*, there is an iron-responsive element (IRE) motif (5′-CTATGT-CAGAG-3′) between bases 1688–1698 (Figure 1a).

A search of the Soybean TIGR Gene Index (http://www.tigr.org) yielded several soybean sequences similar to *GmDmt1;1*. These sequences consisted of expressed sequence tags (ESTs) aligned to make four tentative consensus sequences (TC84846, TC93163, TC94978 and TC82594), while a fifth sequence was from GenBank (accession no. AW277420). These partial sequences are between 65 and 98%, identical to *GmDmt1;1*. Sequence TC93163 has 98% identity with *GmDmt1;1* (isolated from cv. Stevens) and is likely to represent the same isoform from soybean cv. Williams. Obviously, *GmDmt1;1* is a member of a small gene family in soybean.

Gene expression

Northern blot analysis demonstrated that GmDmt1;1 is a nodule-enhanced protein. *GmDmt1;1* mRNA transcripts were abundant in nodules, but were only weakly detected in roots, leaves and stems (Figure 2a). Coincidently, nodule *GmDmt1;1* mRNA expression was the highest during the growth period, associated with maximum rates of symbiotic nitrogen fixation (20–40 days after planting), and decreased thereafter (Figure 2b,c). In young developing nodules, *GmDmt1;1* mRNA was barely detectable (Figure 2b).

Protein localisation

Antibodies were raised in rabbits against the N-terminal 73 amino acids of GmDmt1;1 (Figure 1c). This antiserum was used in Western blot analysis of 4-week-old total soluble nodule proteins, nodule microsomes, PBS proteins and PBM, isolated from purified symbiosomes. The anti-GmDMT1 antiserum identified a 67-kDa protein on the PBM-enriched nodule protein fraction (Figure 3a), but did

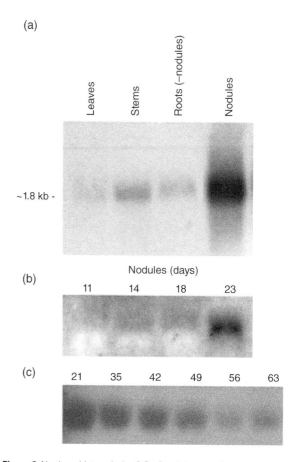

Figure 2 Northern blot analysis of *GmDmt1;1* expression.
(a) *GmDmt1;1* tissue expression. One microgram of poly(A)+-enriched RNA was extracted from 4-week-old soybean leaves, stems, roots (nodules detached) and nodules.
(b) GmDmt1;1 expression in developing nodules.
(c) GmDmt1;1 expression in mature nodules.
Ten micrograms of total RNA was extracted from the nodules prior to and after the onset of symbiotic nitrogen fixation. Blots (a) and (c) were probed with DIG-labelled antisense *GmDmt1;1* full-length RNA, while blot (b) was probed with randomly primed DIG-labelled full-length *GmDmt1;1* cDNA.

not cross-react with soluble nodule proteins, PBS proteins or nodule microsomes (Figure 3a). Replicate Western blots incubated with pre-immune serum (Figure 3b) did not cross-react with the soybean nodule tissues examined. The protein identified on the PBM-enriched protein fraction is approximately 10 kDa larger than that predicted by the amino acid sequence of GmDmt1. The increase in size may be related to extensive post-translational modification (e.g. glycosylation) of GmDmt1, as it occurs in other systems. For example, the human Nramp1 and Nramp2 homologues are extensively modified by glycosylation and can appear about 40% larger on SDS–PAGE than predicted by their amino acid sequence alone (Gruenheid *et al.,* 1999; Tabuchi *et al.,* 2000, 2002). Post-translational modification of PBM proteins has been observed previously (Cheon *et al.,* 1994; Kaiser *et al.,* 1998), and the PBM protein

Nod 24 undergoes extensive post-translation modification en route to the PBM, changing its apparent size on SDS–PAGE from 15 to 32 kDa (Cheon *et al.,* 1994). The localisation of GmDmt1;1 to the PBM was confirmed by subsequent immunogold-labelling experiments on fixed sections of infected cells containing symbiosomes. The anti-GmDmt1;1 antisera cross-reacted primarily with proteins on the PBM (Figure 3c,d). Occasional cross-reactivity with bacteroids was also evident, but this was significantly reduced with more stringent blocking buffers, which included 5% w/v foetal albumin and 3% w/v normal goat serum (Figure 3e).

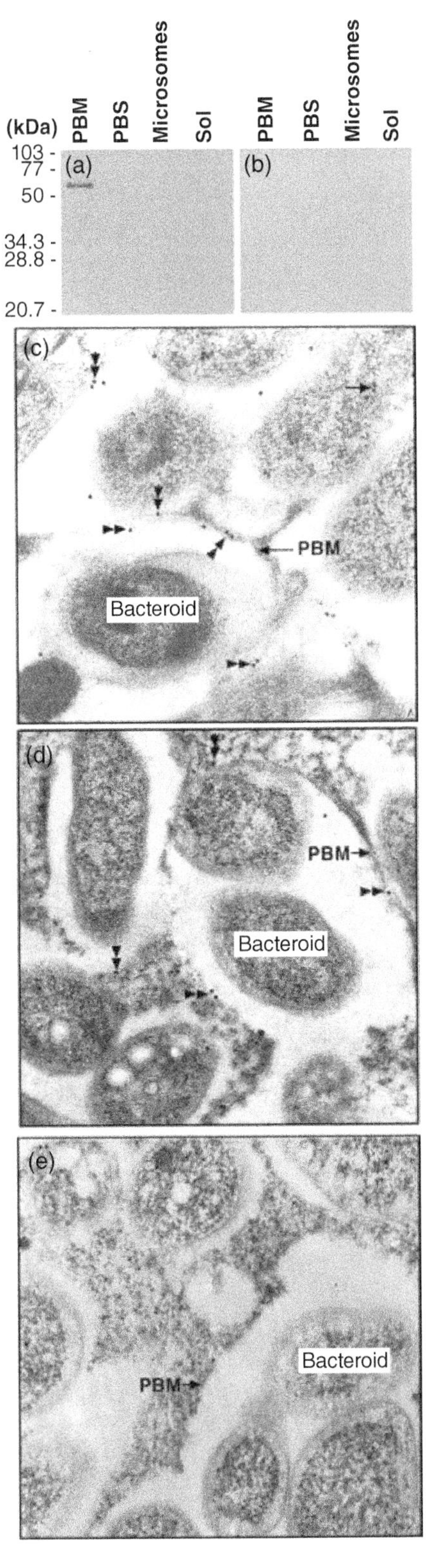

Functional analysis in yeast

To test for Fe^{2+}-transport activity, GmDmt1;1 and the positive control AtIrt1 (a known iron transporter) was cloned into the yeast-expression vectors, pFL61 and pDR195, and then transformed into the yeast iron-transport mutant DEY1453 *(fet3fet4),* which grows poorly on media containing low iron concentrations as a result of disrupted high *(fet3)*- and low (fet4)-affinity Fe^{2+}-transport activity (Dix *et al.,* 1994; Eide *et al.,* 1992). On synthetic-defined (SD) media supplemented with or without 2 μM $FeCl_3$, both AtIrt1 and GmDmt1;1 improved the growth of *fet3fet4* cells over those containing the empty cloning vector pFL61 (Figure 4a). Similarly, in liquid SD media supplemented with 20 μM $FeCl_3$ cells containing either AtIrt1 or GmDmt1;1 routinely entered the exponential-growth phase earlier than those of the empty vector controls (Figure 4b). In the absence of any added iron, GmDmt1;1 was unable to enhance growth of the mutant yeast (results not shown).

Short-term uptake experiments with 1 μM $^{55}FeCl_3$ showed that transformation of *fet3fet4* cells with GmDmt1;1 enhanced accumulation of $^{55}Fe(II)$ approximately fourfold over control cells (Figure 5a). This uptake followed Michaelis–Menten kinetics with an apparent K_M of 6.4 ± 1.1 μM (Figure 5b). The apparent K_M for Fe(II) agrees well with the need for supplementation of growth medium with micromolar iron in order to observe enhanced growth by the GmDmt1;1 cells (see above).

We tested whether GmDmt1;1 can transport other metal ions by heterologous expression in the zinc-deficient yeast-transport mutant, ZHY3 *(zrt1zrt2)* and the

Figure 3 Immunolocalisation of GmDmt1;1 to the peribacteroid membrane (PBM) of soybean nodules.
Western analysis of SDS–PAGE separated and blotted 4-week-old nodule protein fractions including enriched PBM, peribacteroid space (PBS) proteins, total microsomes and soluble proteins. Duplicate blots were incubated with anti-GmDmt1;1 antiserum (a) or with pre-immune antisera (b) at a dilution of 1 : 3000, respectively. Thirty micrograms of purified protein was loaded in each lane. Molecular size markers are shown on the left. (c–e) Immunogold labelling of 3-week-old soybean nodule cross-sections of infected cells with symbiosomes. Tissue sections were incubated with anti-GmDmt1 antisera at a dilution of 1 : 100 (c, d) or with the preimmune serum at a dilution of 1 : 50 (e) followed by 15-nm colloidal gold conjugated with goat antirabbit IgG (BIOCELL EM GAR 15) at a dilution of 1 : 40. Double arrows indicate immunoreactive proteins on the PBM and single arrows identify possible cross-contamination with bacteroids. EM magnification for both pictures was 35 000×.

manganese transport mutant SMF1 (Chen *et al.,* 1999). On minimal zinc plates, GmDmt1 partially complemented ZHY3, but the growth of this mutant was slower

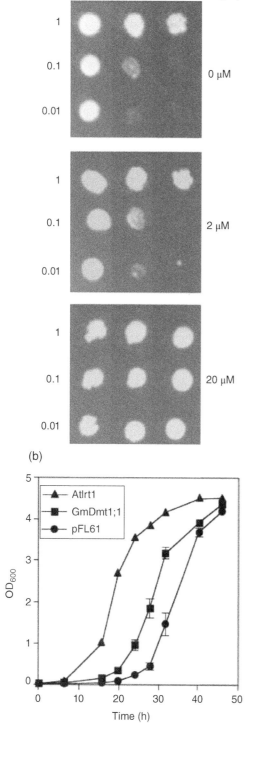

(a)

(b)

than that of DEY1453 *(fet3fet4)* transformed with GmDmt1;1 (mean doubling times were 6.3 ± 0.5 h versus 5.1 ± 0.01 h ($n = 4$), respectively). In short-term transport studies, a 10-fold excess of $MnCl_2$ in the reaction medium inhibited ^{55}Fe uptake significantly by DEY1453 *(fet3fet4)* transformed with GmDmt1;1 (Figure 5c). Similar inhibitions were seen with 10-fold $CuCl_2$ and $ZnCl_2$ (Figure 5c).

Discussion

GmDmt1;1 can transport ferrous iron

The results presented here demonstrate that GmDmt1;1 is a symbiotically enhanced homologue of the Nramp family of divalent metal ion transporters. The sequence of *GmDmt1;1* shares several common features with other members of the family, including 11–12 predicted transmembrane domains, a consensus transport motif between transmembrane domains 8 and 9 and an IRE in the 3′-UTR of the transcript (Gunshin *et al.,* 1997). Its expression is strongly enhanced in nodules, and immunological studies clearly localise the protein to the symbiosome membrane of infected cells. Its ability to rescue growth of the *fet3fet4* yeast mutant on low iron medium makes GmDmt1;1 a strong candidate for the ferrous iron transporter, previously identified in isolated symbiosomes from soybean (Moreau *et al.,* 1998). The kinetics of $^{55}Fe^{2+}$ uptake into complemented yeast (with an apparent K_M of 6.4 μM) also resemble those observed in isolated symbiosomes (linear uptake was observed over the range of 5–50 μM iron; Moreau *et al.,* 1998).

Specificity of GmDmt1;1

The competition experiments shown in Figure 5(c) indicate that GmDmt1 can transport other divalent cations in addition to ferrous iron. Zinc, copper and manganese all inhibited iron uptake. The ability of GmDmt1;1 to enhance growth of the *zrt1zrt2* yeast mutant further suggests that the protein is not specific for iron transport. The preferred substrate *in vivo* may well depend on the relative concentrations of divalent metals in the infected cell cytosol.

Figure 4 Functional analysis of GmDmt1;1 activity in yeast cells. *fet3fet4* yeast cells were transformed with GmDmt1;1 inserted in the expression vector pFL61. Cells were also transformed with empty yeast expression vectors.
(a) Growth of serially diluted cells after 6 days at 30°C of GmDmt1;1 (GmDmt1;1-pFL61), Atlrt1 (Atlrt1-pFL61) and control (pFL61) transformed *fet3fet4* cells on synthetic-defined (SD) media supplemented with 0, 2, 20 μM FeCl3.
(b) Growth in liquid SD media supplemented with 20 μM $FeCl_3$.

This lack of specificity has been found with Nramp homologues from other organisms, including Nramp2 from mice. Despite this lack of specificity when expressed in heterologous systems, mutation of murine Nramp2 results in an anaemic phenotype, demonstrating that *in vivo* it is predominantly an iron transporter (Fleming *et al.,* 1997). Although GmDmt1;1 was able to complement the DEY1453 *(fet3fet4)* yeast mutant, the complementation was not robust and the growth media had to be supplemented with low concentrations of iron. AtIrt1, on the other hand, showed much better complementation and allowed growth of the mutant in the absence of added iron (Figure 4). There are several possible reasons for the poorer growth with GmDmt1;1, including possible instability of GmDmt1;1 transcripts (perhaps because of the presence of the regulatory IRE element in the transcript).

Localisation and function of GmDmt1;1

It has been suggested that AtNramp has an intracellular localisation (Grotz and Guerinot, 2002). The symbiosome is a vacuole-like structure (Mellor, 1989) and contains high concentrations of non-heme iron (Wittenberg *et al.,* 1996). However, this raises an interesting question as to the mechanism of GmDmt1;1. Divalent metal transport into vacuoles is likely to occur as Fe^{2+}/H^+ exchange (Gonzalez *et al.,* 1999), and it is possible that this also occurs in symbiosomes, as the PBM is energised by a H^+-pumping ATPase, which generates a membrane potential positive on the inside (and an acidic interior if permeant anions are present; Udvardi and Day, 1997). However, in this situation, and also in yeast, GmDmt1;1 catalyses uptake of iron into the cell, while uptake into symbiosomes is equivalent to export from the plant cytosol. Assuming that GmDmt1;1 is located in the plasma membrane of yeast and that it has the same physical orientation as in symbiosomes, which is likely considering that the secretory pathway is thought to mediate protein insertion into the PBM, then GmDmt1;1 must be able to catalyse bidirectional transport of iron. This is not unusual for a carrier and has been observed

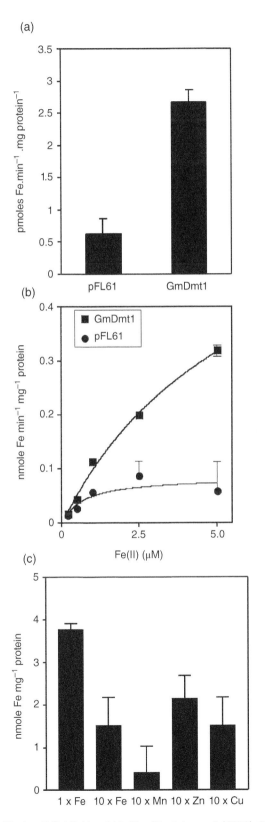

(a)

(b)

(c)

Figure 5 Uptake of Fe(II) by GmDmt1 in yeast.
(a) Influx of $^{55}Fe^{2+}$ into yeast cells transformed with GmDmt1;1. *fet3fet4* cells were transformed with GmDmt1;1-pFL61 or pFL61 and then incubated with 1 μM $^{55}FeCl_3$ (pH 5.5) for 5- and 10-min periods. Data presented are means ± SE of ^{55}Fe uptake between 5 and 10 min from three separate experiments (each performed in triplicate).
(b) Concentration dependence of ^{55}Fe influx into *fet3fet4* cells transformed with GmDmt1;1-pFL61 or pFL61. Data presented are means ± SE of ^{55}Fe uptake over 5 min ($n = 3$). The curve was obtained by direct fit to the Michaelis–Menten equation. Estimated K_M and V_{MAX} for GmDmt1;1 were 6.4 ± 1.1 μM Fe(II) and 0.72 ± 0.08 nM Fe(II) min^{-1} mg^{-1} protein, respectively.
(c) Effect of other divalent cations on uptake of $^{55}Fe^{2+}$ into *fet3fet4* cells transformed with pFL61-GmDMT1;1. Data presented are means ± SE of ^{55}Fe (10 μM) uptake over 10 min in the presence and absence of 100 μM unlabelled Fe^{2+}, Cu^{2+}, Zn^{2+} and Mn^{2+}.

with GmZip1, a zinc transporter on the PBM. It appears that iron uptake can be linked to the membrane potential or pH gradient via other ion movements in the heterologous system. Further experiments on symbiosomes and yeast (or *Xenopus* oocytes) may provide new insights into the mechanism of iron transport in plants, but it appears that GmDmt1;1 has the capacity to function *in vivo* as either an uptake or an efflux mechanism in symbiosomes. This also raises the question of the relationship between GmDmt1;1 and the NADH-ferric chelate reductase on the PBM (Levier and Guerinot, 1996).

At the plant plasma membrane, ferrous iron transporters (presumably AtIrt1 homologues) acttotake up iron reduced by the reductase into the plant. In the symbiosome, assuming that the orientation of the reductase on the PBM is similar to that on the plasma membrane, ferric iron stored in the symbiosome space would be reduced upon oxidation of NADH in the plant cytosol. In isolated symbiosomes, addition of NADH together with ferric citrate, stimulated iron accumulation in the bacteroid, suggesting that the ferrous iron produced in the symbiosome space was taken up by the bacteroid ferrous iron transporter (Moreau *et al.*, 1998). *In vivo*, however, Fe(II) in the symbiosome space could also be transported back into the plant cytosol by the action of GmDmt1;1. We attempted to demonstrate this with isolated symbiosomes by loading them with $^{55}Fe^{3+}$ citrate, adding NADH and ATP (the latter to energise the membrane), and measuring efflux of ^{55}Fe into the reaction medium, but could not detect any efflux (Thomson, data not shown). The direction of transport *in vivo* will depend on the concentration of other ions on either side of the PBM and the activity of the bacteroid ferric and ferrous transporters.

Regulation of GmDmt1;1 expression

As mentioned above, GmDmt1;1 contains an IRE in its 3'-UTR. IREs are conserved sequences in the UTR of certain RNA transcripts to which iron-regulating proteins (IRPs) bind. The presence of an IRE motif suggests that *GmDmt1;1* mRNA may be stabilised by the binding of IRPs in soybean nodules when free iron levels are low. In both mammals (Canonne-Hergaux *et al.*, 1999) and *Arabidopsis* (Curie *et al.*, 2000; Thomine *et al.*, 2000), the abundance of Dmt isoforms containing an IRE element is enhanced by iron deficiency. Iron is required for both plant and bacterial enzymes during nodule development and in the functioning of the mature nodule. *GmDmt1;1* transcripts were detectable in relatively young (11-day-old) nodules and increased as the nodules matured (Figure 2). It is possible that during this time, when the

bacteroid and plant iron requirements are relatively high, free iron levels are low and *GmDMT1* transcripts are stabilised by IRPs. This process could ensure nodule iron transport capacity through increased expression and activity of GmDMT1.

Conclusion

We have identified an Nramp homologue, GmDmt1, which is expressed in soybean nodules and encodes a divalent metal ion transporter located on the symbiosome membrane. The ability of this protein to transport ferrous iron makes it a candidate for the ferrous transport activity previously demonstrated in isolated symbiosomes (Moreau *et al.*, 1998).

Experimental procedures

Plant growth

Soybean (*Glycine max* L. cv. Stevens) seeds were inoculated at planting with *Bradyrhizobium japonicum* USDA 110 and grown in river sand in either glass houses under ambient light between 20 and 30°C, or in controlled-temperature growth rooms at 25°C day and 21°C night temperatures. Plants in the growth chambers were provided with a scheduled (14-h day/10-h night) artificial light (approximately 300 photosynthetic active radiation (PAR) at pot level) period. Plants were irrigated daily with a nutrient solution lacking nitrogen (Delves *et al.*, 1986).

Isolation of GmDmt1;1

Poly(A)+ mRNA was extracted from 6-week-old nodules (Kaiser *et al.*, 1998) and was used to synthesise an adaptor-ligated RACE cDNA library (Clontech; Marathon, Roche, Australia). A 480-bp cDNA amplicon was identified fortuitously from a 5'-RACE PCR experiment using an adaptor-specific primer, AP1: 5'-CCATCC-TAATACGA CTCACTATAGGGC-3' and GmAMTR24: 5'-CGAAC-CAAA GCATGAAGGTCCC-3', a gene-specific primer designed against a partial cDNA of a soybean high-affinity NH_4^+ transporter, GmAMT1 (Kaiser, unpublished results). To amplify the complete *GmDmt1;1* cDNA, PCR experiments were performed using a second 6-week-old nodule cDNA library, which was ligated into the yeast-expression vector pYES3 (Kaiser *et al.*, 1998). Using primers pYES11R: 5'-GCCGCAAATTAAAGCCTTCG-3' and GmDMTF2: 5'-AAGAATAAGGTGCCACCACC-3', a 1.4-kb cDNA was amplified, which included the 3'-terminus of GmDMT1. A full-length clone (1.88 kb) was then subsequently amplified by the PCR from an adaptor-ligated 4-week-old nodule cDNA library (Clontech; Marathon) using high-fidelity Taq DNA polymerase (Roche) and primers AP1 and GmDMT1R21:5′-AAAATTTGAAAGTACTAATACAGAGC-3′. Both strands of the full-length cDNA were sequenced.

Northern analysis

Total RNA was extracted from frozen soybean nodules roots after nodules were detached, stems and leaves using either a Phenol/ Guanidine extraction method (Kaiser *et al.,* 1998) or the Qiagen RNAeasy system (Qiagen, Australia). Poly(A)+ RNA was isolated from total RNA pools using Oligotex resin (Qiagen). Ten micrograms of total RNA or 1 μg of Poly(A)+-enriched RNA was size-separated on a denaturing 1X MOPS 1.2% (w/v) agarose gel containing formaldehyde (Sambrook *et al.,* 1989) and blotted overnight onto Hybond N+ nylon membrane in 20× SSC. RNA was fixed to the membrane by baking at 120°C for 30 min. Blots were hybridised with either a full-length DIG-labelled antisense *GmDmt1;1* RNA produced using the SP6/T7 RNADIG-labelling kit (Roche) or full-length randomly primed DIG-labelled *GmDmt1;1* cDNA. Blots were hybridised overnight at 68°C in DIG-easy hybridisation buffer (Roche). After hybridisation, the blots were washed twice for 15 min in 2× SSC, 1% SDS at ambient temperature, twice at 68°C for 30 min in 0.1× SSC, 1% SDS and twice for 15 min at ambient temperature in 0.1× SSC, 0.1% SDS, followed by chemiluminescent detection of the digoxygenin label using CDP-STAR (Roche).

Antibody generation and Western immunoblot analysis

To generate an antibody to GmDmt1;1, a 236-bp DNA fragment coding for 79 N-terminal amino acids was amplified using the PCR, using primers 5′-TGGCTCGAGCCACCAAGAGCAGCCACT-3′ and 5′-ACCCGAATTCCTGAAGGTCCCCCTCTAAG-3′. The DNA fragment was cloned into pGEMT (Promega, Madison, WI, USA) and was sequenced. The N-terminal DNA fragment was then subcloned into pTrcHisB (Invitrogen, San Diego, CA, USA) in-frame with the Histidine$_{(6)}$-tag and the initiation and termination codon. The resulting construct, pHISDMT1, was transformed into *Escherichia coli* TOP 10F′ cells (Invitrogen) and grown in 500 ml of liquid Solution B (SOB) media containing 50 μg ml^{-1} ampicillin at 37°C to an OD$_{600}$ of 0.5. Expression of the His$_{(6)}$-tag GmDmt1;1 fusion protein was then induced by adding 1 mM isopropyl β-D-thiogalactopyranoside (IPTG) and incubating further for 3 h. Cells were collected and lysed in buffer (8 M urea, 50 mM NaH$_2$PO$_4$, 300 mM NaCl, 1.5 mM imidazole pH 8.0) and disrupted by six cycles of freezing and thawing followed by repeated passage through an 18-gauge needle. Insoluble proteins and cell debris were removed by centrifugation

for 10 min at 16 000 **g**, and the supernatant was collected. The His$_{(6)}$-tagged GmDmt1;1 fusion protein was purified by immobilised metal affinity chromatography (Clontech, San Diego, CA, USA). Eluted protein was concentrated by tricholoracetic acid precipitation and re-suspended in elution buffer containing 8 M urea. The concentrated fusion protein (approximately 200 μg) was mixed with an equal volume of complete Freunds adjuvant (Sigma, USA) and injected into New Zealand White rabbits followed by four subsequent 200-μg injections at 1-month intervals. Ten days after the final injection, crude serum was collected. Protein fractions for Western immunoblot analysis were separated by 12 or 15% w/v SDS–PAGE (Laemmli, 1970) and blotted onto Polyvinylidene Fluoride (PVDF) membranes (Amersham, Buckinghamshire, UK), using a wet-blotting system (Bio-Rad, Regents Park, Australia). Membranes were probed with antiserum to GmDmt1;1 at a dilution of 1 : 3000 in PBS buffer, followed by secondary probing with a horseradish peroxidase-conjugated anti-rabbit IgG antibody. Immunoreactive proteins were visualised by chemiluminescence using a commercial kit (Roche, Australia).

Symbiosome isolation and nodule membrane purification

Symbiosomes were purified from soybean nodule extracts as described before (Day *et al.,* 1989), using a 3-step Percoll gradient. PBM-enriched membrane fractions were purified by rapid vortexing (4 min) of symbiosomes in buffer (350 mM mannitol, 25 mM MES-KOH (pH 7.0), 3 mM MgSO$_4$, 1 mM PMSF; 1 mM pAB; 10 μM E64; 1 mM DTT), followed by centrifugation at 10 000 **g** for 10 min in a SS34 rotor (4°C). The supernatant was collected and centrifuged further at 125 000 **g** for 60 min to separate the PBS proteins from the insoluble PBM-enriched membrane fraction. The PBM pellet was phenol-extracted (Hurkman and Tanaka, 1986), and the PBM and PBS fractions were concentrated by ammonium acetate/ methanol precipitation and re-suspended at room temperature in loading buffer (125 mM Tris pH 6.8, 4% w/v SDS, 20% v/v glycerol, 50 mM DTT, 20% v/v mercaptoethanol, 0.001% w/v bromophenol blue). Soluble and insoluble nodule fractions were prepared by grinding nodules in buffer (25 mM MES-KOH pH 7.0, 350 mM mannitol, 3 mM MgSO$_4$, 1 mM PMSF, 1 mM pAB; 10 μM E64), followed by filtration through four layers of miracloth (Calbiochem, San Diego, CA, USA), and were centrifuged at 10 000 **g**, 4°C for 15 min to separate the bacteroids from the plant fraction. The supernatant was centrifuged further at 125 000 **g**, 4°C for 1 h. The supernatant was collected and concentrated by ammonium acetate/methanol precipitation. The nodule total membrane pellet and soluble protein fractions were re-suspended in loading buffer as described above.

Functional expression in yeast

GmDmt1;1 was cloned into the *NotI* site of theyeast–E. *coli* shuttle vector pDR195 downstream of the P-type ATPase promoter PMA1 (Thomine *et al.,* 2000) or into pFL61 under the control of the phosphoglycerate kinase promoter (Minet *et al.,* 1992). Yeast strain DEY1453 *(fet3fet4)* (Eide *et al.,* 1996) *(MATa/MATα ade2/-+can1/can1 his3/his3 leu2/ leu2 trp1/trp1 ura3/ura3 fet3-2::HIS3/fet3-2::HIS3/fet4-1::LEU2/fet4-1::LEU2)* was transformed (Gietz *et al.,* 1992) and selected for growth on SD media containing 20 mg ml^{-1} glucose and appropriate autotrophic requirements (pH 4.5; Dubois and Grenson, 1979). The media was also supplemented with 10 µM FeCl$_3$ to aid in the growth of *fet3fet4*. Yeast-uptake experiments were performed based on the protocol of Eide *et al.* (1992). fet3fet4 cells transformed with expression plasmids were grown to log phase in SD media with 2 µM additional FeCl$_3$. Log-phase cells were harvested, washed in H$_2$O and diluted in new SD media to an OD$_{600}$ of 0.3 and grown for a further 4 h. Cells were harvested and washed twice with cold MES Glucose Nitriso-acetic acid (MGN) uptake buffer (10 mM MES, pH 5.5, 2% (w/v) glucose, 1 mM nitrilotriacetic acid). Cells were equilibrated at 30°C for 10 min before addition of an equal volume of ^{55}Fe^{2+} solution (MGN buffer, with 10 µM FeCl$_3$, ^{55}FeCl$_3$ and 200 µM ascorbic acid to ensure that iron is in the ferrous form). Cells were incubated at 30°C, and aliquots were taken, filtered and washed five times with 500-µl ice-cold synthetic seawater medium (SSW) (1 mM EDTA, 20 mM trisodium citrate, 1 mM KH$_2$PO$_4$, 1 mM CaCl$_2$, 5 mM MgSO$_4$, 1 mM NaCl (pH 4.2)). Duplicate experiments were performed on ice as a background control for iron binding to cellular material. Internalised ^{55}Fe^{2+} was determined by liquid scintillation counting of the filters. Protein amounts were determined using a modified Lowry assay (Peterson, 1977).

Acknowledgements

This research was financially supported by a grant from the Australian Research Council (D.A. Day), the CNRS Programme International de Cooperation Scientifque, Program 637 (S. Moreau, A. Puppo) and a Canadian National Science and Engineering Research Council Postdoctoral fellowship (B.N. Kaiser). We thank Ghislaine Van de Sype for expert technical assistance with the microscopy.

References

Alonso, J.M., Hirayama, T., Roman, G., Nourizadeh, S. **and** Ecker, J.R. (1999) EIN2, a bifunctional transducer of ethylene and stress responses in *Arabidopsis*. Science, 284, 2148–2152.

Appleby, C.A. (1984) Leghemoglobin and rhizobium respiration. Annu. Rev. Plant Physiol. 35, 443–478.

Belouchi, A., Kwan, T. **and** Gros, P. (1997) Cloning and characterization of the OsNramp family from *Oryza sativa,* a new family of membrane proteins possibly implicated in the transport of metal ions. Plant Mol. Biol. 33, 1085–1092.

Canonne-Hergaux, F., Gruenheid, S., Govoni, G. **and** Gros, P. (1999) The Nramp1 protein and its role in resistance to infection and macrophage function. Proc. Assoc. Am. Physicians, 111, 283–289.

Chen, X.Z., Peng, J.B., Cohen, A., Nelson, H., Nelson, N. **and** Hediger, M.A. (1999) Yeast SMF1 mediates H$^+$-coupled iron uptake with concomitant uncoupled cation currents. J. Biol. Chem. 274, 35089–35094.

Cheon, C., Hong, Z. **and** Verma, D.P.S. (1994) Nodulin-24 follows a novel pathway for integration into the peribacteroid membrane in soybean root nodules. J. Biol. Chem. 269 (9), 6598–6602.

Curie, C., Alonso, J.M., Le Jean, M., Ecker, J.R. **and** Briat, J.F. (2000) Involvement of NRAMP1 from *Arabidopsis thaliana* in iron transport. Biochem. J. 347, 749–755.

Day, D.A., Price, G.D. **and** Udvardi, M.K. (1989) Membrane interface of the *Bradyrhizobium japonicum–Glycine max* symbiosis: peribacteroid units from soybean nodules. Aust. J. Plant Physiol. 16, 69–84.

Day, D.A., Kaiser, B.N., Thomson, R., Udvardi, M.K., Moreau, S. **and** Puppo, A. (2001) Nutrient transport across symbiotic membranes from legume nodules. Aust. J. Plant Physiol. 28, 667–674.

Delves, A.C., Matthews, A., Day, D.A., Carter, A.S., Carroll, B.J. **and** Gresshoff, P.M. (1986) Regulation of the soybean – rhizobium nodule symbiosis by shoot and root factors. Plant Physiol. 82, 588–590.

Dix, D.R., Bridgham, J.T., Broderius, M.A., Byersdorfer, C.A. **and** Eide, D.J. (1994) The *fet4* gene encodes the low-affinity Fe(II) transport protein of *Saccharomyces cerevisiae*. J. Biol. Chem. 269, 26092–26099.

Dubois, E. **and** Grenson, M. (1979) Methylamine/ammonium uptake systems in *Saccharomyces cerevisiae:* multiplicity and regulation. Mol. Gen. Genet. 175, 67–76.

Eide, D., Davis-Kaplan, S., Jordan, I., Sipe, D. and Kaplan, J. (1992) Regulation of iron uptake in Saccharomyces cerevisiae. J. Biol. Chem. 267, 20774–20781.

Eide, D., Broderius, M., Fett, J. **and** Guerinot, M.L. (1996) A novel iron-regulated metal transporter from plants identified by functional expression in yeast. Proc. Natl. Acad. Sci. USA, 93, 5624–5628.

Fleming, M.D., Trenor, C.C., Su, M.A., Foernzler, D., Beier, D.R., Dietrich, W.F. and Andrews, N.C. (1997) Microcytic anaemia mice have a mutation in Nramp2, a candidate iron transporter gene. Nat. Genet. 16, 383–386.

Gietz, D., StJean, A., Woods, R.A. **and** Schiestl, R.H. (1992) Improved method for high-efficiency transformation of intact yeast cells. Nucl. Acids Res. 20, 1425–1420.

Gonzalez, A., Koren'kov, V. **and** Wagner, G.J. (1999) A comparison of Zn, Mn, Cd and Ca-transport mechanisms in oat root tonoplast vesicles. Physiologia Plantarum, 106, 203–209.

Grotz, N. **and** Guerinot, ML. (2002) Limiting nutrients: an old problem with new solutions? Curr. Opin. Plant Biol. 5, 158–163.

Gruenheid, S., Canonne-Hergaux, F., Gauthier, S., Hackam, D.J., Grinstein, S. **and** Gros, P. (1999) The iron-transport protein NRAMP2 is an integral membrane glycoprotein that colocalizes with transferrin in recycling endosomes. J. Exp. Med. 189, 831–841.

Gunshin, H., Mackenzie, B., Berger, U.V., Gunshin, Y., Romero, M.F., Boron, W.F., Nussberger, S., Gollan, J.L. **and** Hediger, M.A. (1997) Cloning and characterization of a mammalian proton-coupled metal ion transporter. Nature, 388, 482–488.

Hurkman, W.J. **and** Tanaka, C.K. (1986) Solubilization of plant-membrane proteins for analysis by two-dimensional gel electrophoresis. Plant Physiol. 81, 802–806.

Kaiser, B.N., Finnegan, P.M., Tyerman, S.D., Whitehead, L.F., Bergersen, F.J., Day, D.A. **and** Udvardi, M.K. (1998) Characterization of an ammonium transport protein from the peribacteroid membrane of soybean nodules. Science, 281, 1202–1206.

Kyte, J. **and** Doolittle, R.F. (1982) A simple method for displaying the hydropathic character of a protein. J. Mol. Biol. 157,105–132.

Laemmli, U.K. (1970) Cleavage of structural proteins during the assembly of the head of bacteriophage T4. Nature, 227,680–685.

Levier, K., Day, D.A. **and** Guerinot, M.L. (1996) Iron uptake by symbiosomes from soybean root nodules. Plant Physiol. 111, 893–900.

Levier, K. **and** Guerinot, M.L. (1996) The *Bradyrhizobium japonicum fega* gene encodes an iron-regulated outer membrane protein with similarity to hydroxamate-type siderophore receptors. J. Bacteriol. 178, 7265–7275.

Mellor, R.B. (1989) Bacteroids in the *Rhizobium-legume* symbiosis inhabit a plant internal lytic compartment: implications for other microbial endosymbioses. J. Exp. Bot. 40, 831–839.

Minet, M., Dufour, M. **and** Lacroute, F. (1992) Complementation of *Saccharomyces cerevisiae* auxotrophix mutants by *Arabidopsis thaliana* cDNAs. Plant J. 2, 417–422.

Moreau, S., Meyer, J.M. **and** Puppo, A. (1995) Uptake of iron by symbiosomes and bacteroids from soybean nodules. FEBS Lett. 361, 225–228.

Moreau, S., Day, D.A. **and** Puppo, A. (1998) Ferrous iron is transported across the peribacteroid membrane of soybean nodules. Planta, 207, 83–87.

Moreau, S., Thomson, R.M., Kaiser, B.N., Trevaskis, B., Guerinot, M.L., Udvardi, M.K., Puppo, A. and Day, D.A.

(2002) GmZIP1 encodes a symbiosis-specific zinc transporter in soybean. J. Biol. Chem. 277, 4738–4746.

Peterson, G.L. (1977) A simplification of the protein assay of Lowry et al. which is more generally applicable. Anal. Biochem. 83, 346–356.

Rodrigues, V., Cheah, P.Y., Ray, K. **and** Chia, W. (1995) Malvolio, the *Drosophila* homologue of mouse Nramp-1 (bcg), is expressed in macrophages and in the nervous system and is required for normal taste behaviour. EMBO J. 14, 3007–3020.

Romheld, V. (1987) Different strategies for iron acquisition in higher plants. Physiol. Plant. 70, 231–234.

Sambrook, J., Fritsch, E.F. **and** Maniatis, T. (1989) *Molecular Cloning: a Laboratory Manual.* Cold Spring Harbour; Cold Spring Harbour Laboratory Press.

Supek, F., Supekova, L., Nelson, H. **and** Nelson, N. (1996) A yeast manganese transporter related to the macrophage protein involved in conferring resistance to mycobacteria. Proc. Natl. Acad. Sci. USA, 93, 5105–5110.

Tabuchi, M., Yoshimori, T., Yamaguchi, K., Yoshida, T. **and** Kishi, F. (2000) Human NRAMP/DMT1, which mediates iron transport across endosomal membranes, is localised to late endosomes and lysosomes in HEp-2 cells. J. Biol. Chem. 275, 22220–22228.

Tabuchi, M., Tanaka, N., Nishida-Kitayama, J., Ohno, H. **and** Kishi, F. (2002) Alternative splicing regulates the subcellular localisation of divalent metal transporter 1 isoforms. Mol. Biol. Cell, 13, 4371–4387.

Tang, C., Robson, A.D. **and** Dilworth, M.J. (1990) A split-root experiment shows that iron is required for nodule initiation in *Lupinus angustifolius* L. New Phytol. 115, 61–67.

Thomine, S., Wang, R.C., Ward, J.M., Crawford, N.M. **and** Schroeder, J.I. (2000) Cadmium and iron transport by members of a plant metal transporter family in *Arabidopsis* with homology to Nramp genes. Proc. Natl. Acad. Sci. USA, 97, 4991–4996.

Udvardi, M.K. **and** Day, D.A. (1997) Metabolite transport across symbiotic membranes of legume nodules. Annu. Rev. Plant Physiol. Plant Mol. Biol. 48, 493–523.

Vert, G., Grotz, N., Dedaldechamp, F., Gaymard, F., Guerinot, M.L., Briata, J.F. **and** Curie, C. (2002) IRT1, an *Arabidopsis* transporter essential for iron uptake from the soil and for plant growth. Plant Cell, 14, 1223–1233.

Vidal, S.M. **and** Gros, P. (1994) Resistance to infection with intracellular parasites – identification of a candidate gene. News Physiol. Sci. 9, 178–183.

Whitehead, L.F. **and** Day, D.A. (1997) The peribacteroid membrane. Physiologia Plantarum, 100, 30–44.

Wittenberg, J.B., Wittenberg, B.A., Day, D.A., Udvardi, M.K. **and** Appleby, C.A. (1996) Siderophore-bound iron in the peribacteroid space of soybean root nodules. Plant Soil, 178, 161–169.

CHAPTER 20

PEA2: Britton-Simmons & Abbott (2008)

Reproduced from Kevin H. Britton-Simmons and Karen C. Abbott. Short- and long-term effects of disturbance and propagule pressure on a biological invasion. *Journal of Ecology* 2008, 96, 68–77 doi: 10.1111/j.1365-2745.2007.01319.x.

Writing Scientific Research Articles: Strategy and Steps, Third Edition.
Margaret Cargill and Patrick O'Connor.
© 2021 John Wiley & Sons Ltd. Published 2021 by John Wiley & Sons Ltd.

Journal of Ecology 2008, **96**, 68–77

doi: 10.1111/j.1365-2745.2007.01319.x

Short- and long-term effects of disturbance and propagule pressure on a biological invasion

Kevin H. Britton-Simmons* and Karen C. Abbott†

Department of Ecology and Evolution, The University of Chicago, 1101 East 57th Street, Chicago, IL, 60637, USA

Summary

1. Invading species typically need to overcome multiple limiting factors simultaneously in order to become established, and understanding how such factors interact to regulate the invasion process remains a major challenge in ecology.

2. We used the invasion of marine algal communities by the seaweed Sargassum muticum as a study system to experimentally investigate the independent and interactive effects of disturbance and propagule pressure in the short term. Based on our experimental results, we parameterized an integrodifference equation model, which we used to examine how disturbances created by different benthic herbivores influence the longer term invasion success of S. muticum.

3. Our experimental results demonstrate that in this system neither disturbance nor propagule input alone was sufficient to maximize invasion success. Rather, the interaction between these processes was critical for understanding how the S. muticum invasion is regulated in the short term.

4. The model showed that both the size and spatial arrangement of herbivore disturbances had a major impact on how disturbance facilitated the invasion, by jointly determining how much space-limitation was alleviated and how readily disturbed areas could be reached by dispersing proagules.

5. *Synthesis*. Both the short-term experiment and the long-term model sow the S. muticum invasion success is co-regulated by disturbance and propagule pressure. Our results underscore the importance of considering interactive effects when making predictions about invasion success.

Key-words: biological invasion, biotic resistance, disturbance, establishment probabitlity, propagule pressure, *Sargassum muticum*

Introduction

Biological invasions are a global problem with substantial economic (Pimentel *et al.* 2005) and ecological (Mack *et al.* 2000) costs. Research on invasions has provided important insights into the establishment, spread and impact of nonnative species. One key goal of invasion biology has been to identify the factors that determine whether an invasion will be successful (Williamson 1996). Accordingly, ecologists have identified several individual factors (e.g. disturbance and propagule pressure) that appear to exert strong controlling influences on the invasion process. However, understanding how these processes interact to regulate invasions remains a major challenge in ecology (D'Antonio *et al.* 2001; Lockwood *et al.* 2005; Von Holle & Simberloff 2005).

Propagule pressure is widely recognized as an important factor that influences invasion success (MacDonald *et al.* 1989; Simberloff 1989; Williamson 1996; Lonsdale 1999; Cassey *et al.* 2005). Previous studies suggest that the probability of a successful invasion increases with the number of propagules released (Panetta & Randall 1994; Williamson 1989; Grevstad 1999), with the number of introduction attempts (Veltman *et al.* 1996), with introduction rate (Drake *et al.* 2005), and with proximity to existing populations of invaders (Bossenbroek *et al.* 2001). Moreover, propagule pressure may influence invasion dynamics after establishment by affecting the capacity of non-native species to adapt to their new environment (Ahlroth *et al.* 2003; Travis *et al.* 2005). Despite its acknowledged importance, propagule pressure has rarely been manipulated experimentally and the interaction of propagule pressure with other processes that regulate invasion success is not well understood (D'Antonio *et al.* 2001; Lockwood *et al.* 2005).

*Correspondence and present address. Friday Harbor Laboratories, University of Washington, 620 University Road, Friday Harbor, WA 98250, USA. E-mail: aquaman@u.washington.edu

†(Present address: Department of Zoology, University of Wisconsin, 430 Lincoln Drive, Madison, WI 53706, USA

© 2007 The Authors. Journal compilation © 2007 British Ecological Society

Resource availability is a second key factor known to influence invasion success and processes that increase or decrease resource availability therefore have strong effects on invasions (Davis *et al.* 2000). Resource pre-emption by native species generates biotic resistance to invasion (Stachowicz *et al.* 1999; Naeem *et al.* 2000; Levine *et al.* 2004). Consequently, physical disturbance can facilitate invasions by reducing competition for limiting resources (Richardson & Bond 1991; Hobbs & Huenneke 1992; Kotanen 1997; Prieur-Richard & Lavorel 2000). In most communities disturbances occur via multiple mechanisms and the disturbances created by different agents vary in their intensity and frequency (D'Antonio *et al.* 1999). Recent empirical (Larson 2003; Hill *et al.* 2005) and theoretical (Higgins & Richardson 1998) studies suggest that not all types of disturbance have equivalent effects on the invasion process. Moreover, most of what we know about the effects of disturbance on invasions comes from short-term experimental studies. It is presently unclear how different disturbance agents influence long-term patterns of invasion.

In order for any invasion to be successful, propagule arrival must coincide with the availability of resources needed by the invading species (Davis *et al.* 2000). Therefore, the interaction between propagule pressure and processes that influence resource availability will ultimately determine invasion success (Brown & Peet 2003; Lockwood *et al.* 2005; Buckley *et al.* 2007). In this study we used the invasion of shallow, subtidal kelp communities in Washington State by the Japanese seaweed *Sargassum muticum* as a study system to better understand the effects of propagule pressure and disturbance on invasion. In a factorial field experiment we manipulated both propagule pressure and disturbance in order to examine how these factors independently and interactively influence *S. muticum* establishment in the short term. We supplement the experimental results with a parameterized integrodifference equation model, which we use to examine how different natural disturbance agents influence the spread of *S. muticum* through the habitat in the longer term. Although a successful invasion clearly requires both establishment and spread of the invader, most studies have looked at just one of these processes (Melbourne *et al.* 2007). We take an integrative approach by employing both a short-term experiment and a longer-term model, allowing us to examine the effects of disturbance and propagule limitation on the entire invasion process.

Methods

STUDY SYSTEM

Our field research was based out of Friday Harbor Laboratories on San Juan Island, Washington State, USA. The field experiment was carried out at a site within the San Juan Islands Marine Preserve network adjacent to Shaw Island, known locally as Point George (48.5549° N, 122.9810° W). Field work was accomplished using SCUBA in shallow subtidal communities.

The native algal community characteristic of sheltered, rocky subtidal habitats in this region is species-rich and structurally complex (see Britton-Simmons 2006 for a more detailed description). In this ecosystem, space is an important limiting resource and in the absence of disturbance there is little or no bare rock available for newly arriving organisms to colonize. This habitat has a diverse fauna of benthic herbivores, including molluscs and sea urchins, that create disturbances by clearing algae from the rocky substrata. The green sea urchin *Strongylocentrotus droebachiensis* is a generalist herbivore that reduces the abundance of native algae and creates relatively large disturbed patches (Vadas 1968; Duggins 1980). In the shallow zone where *S. muticum* is found, the green urchin is highly mobile and often occurs in aggregations (Paine & Vadas 1969; Foreman 1977; Duggins 1983; personal observation). Green urchins avoid areas where *S. muticum* is present because it is not a preferred food resource (Britton-Simmons 2004), but they can be found feeding in uninvaded areas adjacent to existing *S. muticum* populations (personal observation). Green urchins therefore create intermittent but relatively intense disturbances in areas where *S. muticum* is absent and some proportion of these disturbances can potentially be exploited by dispersing *S. muticum* propagules. In contrast, herbivorous benthic molluscs (chitons, limpets and snails) are ubiquitous in the shallow subtidal and unlike sea urchins they are unaffected by the presence of *S. muticum* (Britton-Simmons 2004). Herbivory by individual molluscs creates relatively small-scale disturbances, thereby providing a consistent supply of microsites that can be colonized by newly arriving species, including *Sargassum muticum* (see Appendix S1 in Supplementary Material for more information about mollusc diets).

THE INVADER

Sargassum muticum is a brown alga in the order Fucales that was introduced to Washington State in the early 20th century, probably with shipments of Japanese oysters that were imported for aquaculture beginning in 1902 (Scagel 1956). It is now common in shallow subtidal habitats throughout Puget Sound and the San Juan Islands (Nearshore Habitat Program 2001, personal observation). In the San Juan Islands, *S. muticum* has a pseudoperennial life history. Each holdfast produces as many as 18 laterals in the early spring, each of which can grow as tall as three metres. In late summer to early autumn the laterals senesce and are lost, leaving only the basal holdfast portion of the thallus to overwinter.

Sargassum muticum has a diplontic (uniphasic) life cycle, is monoecious, and is capable of selfing. Reproduction typically occurs between late June and late August in our region. During reproduction the eggs of *S. muticum* are released from and subsequently adhere to the outside of small reproductive structures called receptacles. Once fertilized, the resulting embryos remain attached while they develop into tiny germlings (< 200 μm in length) with adhesive rhizoids (Deysher & Norton 1982). Germlings then detach from the receptacle and sink relatively quickly, recruiting in close proximity to the parent plant (Deysher & Norton 1982). Although most recruitment occurs within 5 m of adult plants, recruits have been found as far as 30 m from the nearest adult (Deysher & Norton 1982). Longer distance dispersal probably occurs when plants get detached from the substratum and subsequently become fertile after drifting for some period of time (Deysher & Norton 1982). One distinctive feature of the *S. muticum* invasion is that it is extremely limited in vertical extent. In the San Juan Islands, *S. muticum* is found from the low intertidal to the shallow subtidal zone (Norton 1977; personal observation), from approximately –0.5 m Mean Lower Low Water (MLLW) to –7 m MLLW. However, it is most abundant in the shallow subtidal, from approximately –2 m MLLW to –4 m MLLW. Thus, in areas where *S. muticum* has invaded it forms a narrow band along the shore.

FIELD EXPERIMENT

We used a two-way factorial design manipulating propagule pressure (six levels) and disturbance (two levels) with three replicates per treatment combination. Subtidal plots (30 cm × 30 cm) at a depth of 3–4 m below MLLW were selected so that differences in the identity and abundance of taxa, aspect, and relief were minimized and the plots were randomly assigned to treatments. None of the experimental plots contained *S. muticum* prior to the experiment. However, some *S. muticum* was present at Point George and it was removed prior to the reproductive season in order to prevent contamination of the experimental plots from external sources of *S. muticum* propagules.

The disturbance treatment had two levels: control and disturbed. Control plots were not altered in any way, but they did vary somewhat in how much natural disturbance had occurred in them prior to the experiment (mean = 7.7% of plot area). Plots in the disturbance treatment were scraped down to bare rock so that no visible organisms remained. These two treatments represent extremes in the levels of disturbance that are likely to occur in nature. The unaltered control plots contained a rich assemblage of native species. The disturbed plots were similar in spatial scale to a patch that a small group of urchins might create, but represent an unusually intense disturbance because all native species, including crustose coralline algae (which cover an average of 27.7% of the substratum at this depth), were removed. These treatments maximized our ability to detect an effect of disturbance in our experiment.

Immediately following the imposition of the disturbance treatment (July 2002) the plots were experimentally invaded by suspending 'brooding' *S. muticum* over them. This was accomplished by collecting *S. muticum* from the field and transporting them to the lab where the appropriate ratio of sterile to reproductive tissue (see below) was placed in 30 cm × 30 cm vexar bags. The bags were returned to the field the same day and suspended over the experimental plots for 1 week. Propagule pressure was manipulated by varying the ratio of sterile to reproductive tissue in the bags while holding the total biomass of *S. muticum* tissue constant. The propagule pressure treatment had six levels, corresponding to the following amounts of reproductive tissue (in grams): 0, 50, 100, 175, 250 and 350 (average mass of mature *S. muticum* in this region is 174 g). Based on propagule production–mass relationships derived by Norton & Deysher (1988) for *S. muticum*, we estimate that approximately 5 million propagules were released in each replicate of our highest propagule pressure treatment. We assumed a linear relationship between the mass of adult reproductive tissue and propagule output because we know of no *Sargassum* study that suggests otherwise. Sterile tissue was added to bags as necessary in order to bring the total biomass to 350 g. Reproductive and sterile tissue was mixed in the bags so that the reproductive tissue was well distributed throughout. This experimental manipulation mimics the level of propagule input that would occur in an incipient invasion or if a drifting plant became tangled with attached algae and subsequently released its propagules.

Recruitment of *S. muticum* was quantified by counting the number of *S. muticum* juveniles that were present in the plots 5 months after the experimental invasion, which is the earliest they can reliably be seen in the field. We resurveyed the plots to count the number of *S. muticum* adults present 11 months after the invasion (just prior to reproductive season) and then removed all *S. muticum* from the experimental plots in order to prevent it from spreading.

STATISTICAL ANALYSIS

We analysed the *S. muticum* recruitment data using a two-way ANOVA followed by separate regression analyses on each disturbance treatment. For the control treatment, we performed a multiple regression to determine what proportion of recruitment variation was explained by propagule input and space availability. For the disturbed plots, which did not vary in the amount of available space, we carried out a simple linear regression to determine the impact of propagule input on recruitment. We used the results of these analyses to inform the construction of mechanistic candidate functions for the relationship between propagule input, space availability and recruitment. These candidate functions were compared using differences in the Akaike's information criteria (AIC differences; Burnham & Anderson 2002). We then used model averaging, a form of multimodel inference in which parameter estimates from more than one candidate function are used jointly to describe the data, in order to select a parameterized recruitment function for the *S. muticum* spread model.

The *S. muticum* survivorship data did not conform to the assumptions of ANOVA (even after a number of different transformations) so we used a non-parametric Kruskal–Wallis test to ask whether *S. muticum* survivorship differed in the disturbed and control treatments. We then fitted five different survivorship functions, assuming binomial error, to the data to test whether *S. muticum* survivorship (number of adults per recruit) was density-dependent. Because the Kruskal–Wallis test suggested that survivorship differed significantly between the two disturbance treatments (see Results) we chose to fit the models to those two treatments separately to test for density dependence. In addition to type 1 (linear), type 2 (saturating), and type 3 (sigmoidal) functions, we also fitted a constant survivorship model. These candidate functions were compared using the Akaike's information criterion (AIC differences; Burnham & Anderson 2002).

The numbers of adult *S. muticum* (after 11 months) also violated the assumptions of ANOVA (despite transformations), so we used non-parametric statistics to test two hypotheses: (i) adult density is independent of disturbance treatment (Wilcoxon Signed Ranks Test), and (ii) adult density is independent of propagule pressure treatment (Kruskal–Wallis Test).

MODEL

We used an integrodifference equation (IDE) model to describe the spatial spread of an S. *muticum* population. IDE models assume that the habitat is continuous in space, and that reproduction and dispersal occur in discrete bouts. The depths inhabited by *S. muticum* comprise a relatively narrow vertical band, so the spread of the population was assumed to occur in a one-dimensional habitat. The model follows two state variables through time. $N_t(x)$ is the density of *S. muticum* at a location x along this habitat at time t, and $Z_t(x)$ is the amount of bare rock at x during t. The values for these state variables are determined by functions representing the important ecological processes in this system. *Sargassum muticum* density is determined by the production and recruitment of propagules and by adult survival. Bare rock is created by benthic herbivore disturbances, since herbivores consume native algae and thus alleviate space limitation. The form of our model is then

$$N_{t+1}(x) = sP_t(x)f\left(P_t(x), Z_t(x)\right) + rN_t(x), \qquad \text{eqn 1}$$

$$Z_{t+1}(x) = \left(1 - \eta_t(x)\right)gZ_t(x) + \eta_t(x)A. \qquad \text{eqn 2}$$

$P_t(x)$ is the number of propagules at location x at the start of year t, and equals the number of propagules produced at x and remaining near their parent plant plus the sum of propagules from all other locations within the habitat (with endpoints a and b) which disperse to x. $P_t(x)$ is governed by the equation $P_t\left(x\right)=\int_a^b \omega N_t\left(y\right)k\left(x-y\right)dy$.

Each adult produces ω propagules and their dispersal is described by the function k. The function $f(P_t(x), Z_t(x))$ in equation 1 gives the fraction of propagules which successfully recruit, given that the amount of bare rock at location x equals $Z_t(x)$ and there is an initial input of $P_t(x)$ propagules. Based on data from the experiment, we assume that recruitment function has the form $f(P_t(x), Z_t(x)) = \rho_1(Z_t(x) + \rho_2)^{\rho 5} P_t(x)/[1 + \rho_3(Z_t(x) + \rho_2)^{\rho 5} + \rho_4 P_t(x)^2]$, with values for the ρ_i and methods for fitting this function given in Appendix S2. s and r are fractions of germlings and adults, respectively, that survive to the following year. Parameters for *Sargassum* fecundity and dispersal were attained from the literature (Deysher & Norton 1982; Norton & Deysher 1988) and all other parameter values used in our simulations were estimated from our own field data. The methods and results for fitting parameters are given in Appendix S2.

In equation 2, $\eta_t(x)$ is the proportion of the habitat scraped clear by grazers. If left ungrazed, we assumed that bare rock at a given location experiences geometric decay, with rate g, as it becomes utilized by native algae. The parameter A in equation 2 is a scaling constant representing the size of the habitable area at each point x. We modelled benthic herbivore disturbance in two different ways. First, we constructed a stochastic model for $\eta_t(x)$ based on our understanding of the natural history of the system. Second, we built a more generalized stochastic model for $\eta_t(x)$. In the *S. muticum* system, bare rock is generated in small patches when an area is grazed by molluscs (chitons and limpets), or in larger patches by sea urchin grazing. Both types of disturbance create bare rock for *S. muticum* to potentially exploit, and the disturbance types differ only in their size and spatial distribution. We assumed that the mollusc disturbances are ubiquitous, whereas large urchin-grazed areas are patchily distributed across the habitat. Due to uncertainty in the exact size and frequency of these disturbances, we ran simulations over a very wide range of possible parameter values. In the generalized model for $\eta_t(x)$, we allowed disturbances of any size to occur with any degree of spatial aggregation, rather than requiring large disturbances to be patchy and small ones to be spread throughout the habitat. Our methods for drawing values for $\eta_t(x)$ in these simulations are described in Appendix S3 and summarized in Table C.1 therein.

In our system, native benthic grazers do not eat *S. muticum* adults (Britton-Simmons 2004; personal observation), but it is unknown whether they will consume new *S. muticum* recruits when they are very small (e.g. Sjøtun *et al.* 2007) and hence difficult to avoid ingesting incidentally. Whether or not disturbance events can directly cause mortality of the invader can be very important in determining invasion success (Buckley *et al.* 2007). In our simulations, we therefore considered both the case where *S. muticum* is never eaten by grazers, and the case where *S. muticum* is eaten at the rate $\eta_t(x)$ until it reaches the age of 1 year.

Results

The field experiment showed that recruitment of *S. muticum* was higher in plots that were disturbed compared to control plots (Fig. 1a) suggesting that resource availability limited recruitment. Increasing propagule pressure led to significant increases in average *S. muticum* recruitment in both distur-

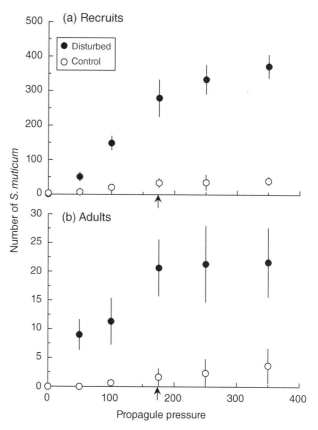

Fig. 1. Number of *Sargassum muticum* (a) recruits and (b) adults in field experiment plots (900 cm²). Propagule pressure is grams of reproductive tissue suspended over experimental plots at beginning of experiment. The average mass of an adult *S. muticum* (174 g) is indicated by an arrow. Data are means ± 1 SE ($n = 3$).

bance treatments (Fig. 1a). Finally, a significant interaction between disturbance and propagule pressure ($F_{5,24} = 3.77$, $P = 0.01$) indicates that the plots in the two disturbance treatments differed in the extent to which they were limited by propagule availability. Multiple regression analysis of the *S. muticum* recruitment data from the control treatment, with space and propagule input as continuous explanatory variables, explained most of the recruitment variability ($R^2 = 0.87$, Fig. 1a). This analysis showed that both space (Fig. 1a, $b = 0.703$, $P < 10^{-4}$) and propagule treatment (Fig. 1a, $b = 0.657$, $P < 10^{-3}$) had strong influences on recruitment in the control treatment. Because there was no variation in space availability in the disturbed treatment, we used simple linear regression analysis to examine the relationship between propagule input and *S. muticum* recruitment in the disturbed treatment (Fig. 1a, $R^2 = 0.84$, $P < 10^{-6}$). The results suggest that in the absence of space limitation propagule input explains most of the variability in *S. muticum* recruitment.

We used these results to create a set of mechanistic candidate functions for the relationship between *S. muticum* recruitment, propagule pressure and space availability (see Appendix S2). The only candidate models supported by the data (AIC differences < 4; Burnham & Anderson 2002) show a type 3 (sigmoidal) relationship between propagule pressure and recruitment, and either a type 2 (saturating) or type 3 relationship between available space and recruitment (Appendix S2, Table B.1). Due to practical constraints on the number of treatments that could be replicated in the field, we have data only on very low available space (control plots) and very high available space (disturbed plots), and insufficient data at intermediate values to resolve the functional relationship between space-limitation and recruitment. We therefore used model averaging (Burnham & Anderson 2002) to combine our parameter estimates for the two supported models and used the resulting function to describe space- and propagule-limitation in recruitment in the simulation model. We also ran simulations using each of the supported recruitment models separately. The results from the two supported models and the averaged model were very similar, so we present results only from the averaged model.

Survivorship (from 5 months to 11 months of age) of *S. muticum* was significantly higher in disturbed plots ($U = 76.5$, $P < 0.05$). Mean survivorship (± 1 SD) in control plots was 3.4% ($\pm 3.8\%$), compared to 6.1% ($\pm 2.2\%$) in disturbed plots. Our analysis of survivorship as a function of recruitment density suggests density-independence (Appendix S2, Table B.2), so we used the mean survivorship across all experimental plots as the germling survival rate (s) in our model.

Simulations of the parameterized model under various disturbance regimes reveal several interesting patterns. Using the disturbance scenario with ubiquitous mollusc disturbances and large, patchily distributed urchin disturbances, we found that a single adult *S. muticum* was almost always sufficient to start a successful invasion. This is in agreement with our empirical observation that propagule input always resulted in positive recruitment, even in space-poor control plots. We quantified population growth in our model by reporting the density of *S. muticum* after 100 years, averaged across the invaded area, and we use the length of habitat occupied by *S. muticum* after 100 years as a measure of invasion rate. When we assumed that *S. muticum* was never consumed by benthic herbivores, both the mean *S. muticum* population density and the length of the invaded area increased with both the mean intensity of mollusc grazing and with the size and number of urchin disturbances (Fig. 2, solid lines). Changing the variance in the intensity of mollusc grazing had essentially no effect (not shown). Unless urchin disturbances were extremely large and numerous (top 3 lines, Fig. 2g–j), the mollusc grazing had a much stronger effect on *S. muticum* density than did urchin grazing.

When we assumed that native grazers eat *S. muticum* germlings, *S. muticum* density and the length of habitat invaded still increased with the intensity of mollusc disturbance, as long as molluscs grazed less than 50% of the habitat bare (Fig. 2, dashed lines). Actual mollusc disturbances are typically much smaller than 50% (personal observation). Indeed, we note that if all of the bare rock in the experiment's control plots was attributed to mollusc grazing, the average grazing intensity would be only 7.7%. Within the realistic range of parameter values, then, molluscs facilitate the invasion in the model even when they consume young *S. muticum*.

Urchin disturbances that were few and/or small had little effect on the invasion, but large and numerous urchin disturbances decreased the final *S. muticum* density and the size of the invaded area when grazers consumed new recruits (Fig. 2e–j). *Sargassum muticum* failed to establish when urchin disturbances were both very large (20–50 m of linear habitat scraped bare per disturbance) and extremely abundant (100–200 such disturbances per year). These results are corroborated by the generalized model of disturbance, which showed that when the total proportion of the habitat disturbed per year is held constant smaller disturbances affecting a greater number of locations resulted in the highest final *S. muticum* densities and invaded areas (Appendix S2, Fig. C.1). When these disturbed locations were more clumped in space, this resulted in a slight decrease in the final size of the invaded area.

The treatment effects were still apparent when adults were counted at the end of the experiment (Fig. 1b). Adult *S. muticum* density was higher in the disturbed treatment than in the control treatment ($Z = -3.41$, $P < 0.001$). In addition, adult *S. muticum* density appeared to be positively related to propagule pressure (Fig. 1b, $H_5 = 16.10$, $P = 0.006$), with high propagule pressure resulting in a maximum of between 20 and 25 adults per plot (900 cm^2).

How was the probability of successful invasion influenced by propagule pressure? We defined successful invasion of an experimental plot as the presence of one or more adult *S. muticum* at the end of the experiment (11 months after invasion). We consider this a reasonable way to define invasion success given that reproduction of these adults was imminent (< 1 month away), survivorship is very high at this life-history stage (Appendix S2, Table B.3), and both our model and experimental results indicate that a single individual is capable of establishing a population. We plotted the proportion of plots in each treatment combination that were successfully invaded as a function of propagule pressure (Fig. 3). Because we had only three replicates per treatment combination the probability values were constrained to four possible values (0, 0.33, 0.66, or 1.0). In addition, we tested only six levels of propagule input and therefore have limited capacity to resolve the details of this relationship. Therefore, we did not attempt to fit statistical models to these data. In disturbed plots, invasion was certain even at the lowest level of propagule pressure in our experiment (Fig. 3). However, in control plots the probability of invasion was less than 1 until propagule pressure reached a level of 250 g of reproductive tissue, an amount of tissue greater than the average mass of an adult *S. muticum* (Fig. 3).

Discussion

Our experimental results demonstrate that space- and propagule-limitation both regulate *S. muticum* recruitment. Our finding that *S. muticum* recruitment was positively related to propagule input is similar to those of two previous studies (Parker 2001; Thomsen *et al.* 2006), in which the propagule input of invasive plants was manipulated. In our control

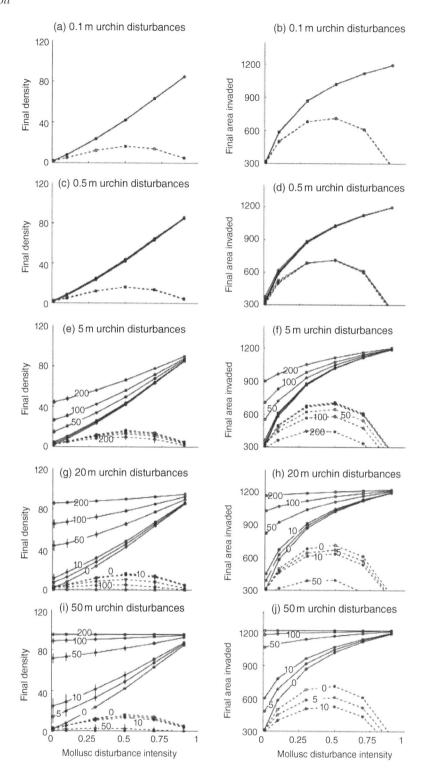

Fig.2 Simulation results using the mollusc/urchin model for disturbance. The first column (a, c, e, g and i) shows the mean *Sargassum muticum* density (individuals per 900 cm²) and the second column (b, d, f, h and j) show the length of habitat occupied (metres) after 100 years. Solid lines are the results when native grazers never eat *S. muticum* and dashed lines are results when *S. muticum* recruits (less than 1 year old) are eaten by grazers. The *x*-axis in all plots shows the average proportion of rock scraped bare by molluscs. The number superimposed on each line is the number of urchin disturbances per year (numbers are omitted when the lines overlap completely or are very close together). The mean size of these urchin disturbances increases from the top row (a–b) to the bottom (i–j) and is printed at the top of each graph. Error bars, when large enough to be visible, are ± 1 SE (*n* = 100, as averages were taken across two values for the variance in mollusc intensity with 50 replicates each).

treatment space was limiting, a result that has also been found in previous studies of *S. muticum* recruitment (Deysher & Norton 1982; De Wreede 1983; Sanchez & Fernandez 2006). Consequently, increasing propagule pressure had a relatively weak effect on recruitment in undisturbed plots (Fig. 1a). However, when space limitation was alleviated by disturbing the plots, increasing propagule pressure caused a dramatic increase in recruitment (Fig. 1a). This suggests that in the presence of adequate substratum for settlement, propagule limita-

tion becomes the primary factor controlling *S. muticum* recruitment. These results indicate that *S. muticum* recruitment under natural field conditions will be determined by the interaction between disturbance and propagule input.

Only a few previous studies have investigated the effect of resource supply on the relationship between propagule pressure and recruitment of an introduced species. Although disturbance generally increases invasion success by increasing resource availability (Richardson & Bond 1991; Bergelson

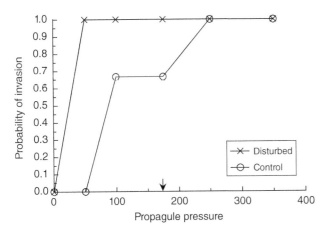

Fig 3 Probability of invasion as a function of propagule pressure. Probability of invasion is the proportion of plots in each treatment combination (*n* = 3) that contained at least one adult *Sargassum muticum* at the end of the experiment. The average mass of an adult *S. muticum* (174 g) is indicated by an arrow.

et al. 1993; Levin *et al.* 2002; Valentine & Johnson 2003; Clark & Johnston 2005), Parker (2001) found evidence that disturbance reduced Scotch broom (*Cystisus scoparius*) recruitment from seed at all levels of propagule input. This effect occurred because the native flora actually facilitated Scotch broom germination, probably by increasing soil moisture and/or nutrients (Parker 2001). Similarly, Thomsen *et al.* (2006) showed that in the absence of a water addition treatment establishment of an exotic perennial grass was greatly reduced, even at high levels of propagule input. Finally, Valentine & Johnson (2003) found that disturbance facilitated invasion by the introduced kelp *Undaria pinnatifida* even when propagule pressure was high. These studies and our own work provide empirical evidence that the interaction between propagule input and the biotic and abiotic processes that mediate resource availability will be key to understanding patterns of invasion.

The effects of the disturbance and propagule pressure treatments that were manifest in the *S. muticum* recruitment data persisted until the end of the experiment (Fig. 1b). That adult *S. muticum* density was higher in the disturbed treatment than in the control treatment suggests that disturbance may increase the population growth rate of *S. muticum* during the initial stages of the invasion. Natural disturbances that are less intense than our experimental scrapings might have a more modest effect on *S. muticum* density, but our simulation results suggest that even small disturbances can play a major role in facilitating the invasion. Our simulations further suggest that this effect should persist over long time-scales (Fig. 2).

In subtidal habitats both biotic and abiotic disturbances occur, but it is doubtful that they are both relevant to the *S. muticum* invasion in this system. Consumption of algae by the diverse fauna of benthic herbivores in this system (see Methods) is a common and consistent source of disturbance

that is likely to be relevant to the *S. muticum* invasion and was therefore the focus of our model. Abiotic disturbances are unlikely to play an important role in this regard because tidal currents are not a substantial cause of algal mortality in this region (Duggins *et al.* 2003) and the inland waters of Puget Sound, the San Juan Islands and the Strait of Georgia are protected from the ocean swells that play a key role on the outer coast of Washington State. Although locally generated storm waves are an important source of disturbance during the winter (Duggins *et al.* 2003), storms during the summer months when *S. muticum* is reproductive are rare.

SIMULATED URCHIN/MOLLUSC DISTURBANCES

In addition to enhancing *S. muticum* recruitment, disturbance increased the survivorship of juvenile *S. muticum*. In our system, the green urchin (*Strongylocentrotus droebachiensis*) creates relatively large disturbed patches and *S. muticum* that recruit to these patches probably benefit from reduced competition with native algae. Unlike other systems where sea urchins feed on both native and non-native algae alike (Valentine & Johnson 2005), green urchins do not consume adult *S. muticum* (Britton-Simmons 2004) although it is possible that they incidentally consume new recruits. Studies in other systems have also reported positive effects of disturbance on the survivorship of non-native species (Gentle & Duggin 1997; Williamson & Harrison 2002). In general, disturbance probably enhances survivorship because it reduces the size or abundance of native species that compete for resources with invaders (Gentle & Duggin 1997; Britton-Simmons 2006). Indeed, our modelling results suggest that even when juvenile survivorship is reduced by herbivory, the net effect of grazers is still usually positive (Fig. 2).

The simulation model suggested that not all disturbance agents have equivalent effects on space-limitation. Small bare patches throughout the habitat facilitated *S. muticum* spread (Fig. 2 and Appendix S3, Fig. C.1) by increasing the amount of bare rock near any given reproductive adult. Molluscs are ubiquitous in these subtidal habitats and although they typically create very small disturbances, the model suggests that this is sufficient for *S. muticum* to successfully invade, even in the absence of other disturbance agents (e.g. urchins and humans).

Urchins create much larger open spaces, but urchin disturbances could not be used by settling propagules unless a reproductive adult happened to be nearby or a long-distance dispersal event occurred. When there are many urchin disturbances in a year, the chance that such a disturbance occurs near an *S. muticum* adult increases and, because long-distance propagule dispersal is rare, this greatly enhances the likelihood that a propagule will reach the disturbed area. Accordingly, small numbers of urchin disturbances in our model did not affect the spread of *S. muticum* (Fig. 2a–d), but numerous and sufficiently large disturbances did (Fig. 2e–j). Washington State is at the southern end of the green urchin's range in the eastern Pacific and at the majority of sites in the San Juan Islands this species is absent or at relatively low abundance. Consequently, molluscs are probably the most important source of disturbance for *S. muticum* in this region; green urchins may be a more important disturbance agent in more northerly portions of its range (where it reaches higher densities). That urchin disturbance was not necessary for successful invasion by *S. muticum* in the model is an important result because *S. muticum* has invaded many areas in this region where urchins are absent. Indeed, urchins avoid areas where *S. muticum* is present (Britton-Simmons 2004) and since this effect was not included in the model, urchin disturbances probably contribute even less to *S. muticum* spread than our simulations suggest.

PROPAGULE PRESSURE AND INVASION SUCCESS

How much invasion risk does a given level of propagule pressure pose? Previous studies have demonstrated a positive relationship between propagule pressure and the establishment success of non-native species (Grevstad 1999; Parker 2001; Ahlroth *et al.* 2003; Cassey *et al.* 2005). However, we know very little about the relationship between establishment probability and propagule pressure or the factors that affect it (Lockwood *et al.* 2005). Possibilities include a linear relationship (Lockwood *et al.* 2005) as well as more complex relationships containing thresholds or other non-linearities (Griffith *et al.* 1989; Ruiz & Carlton 2003; Lockwood *et al.* 2005; Buckley *et al.* 2007). Our experimental results suggest that the relationship is non-linear (Fig. 3). Indeed, all communities in which abiotic factors do not preclude invasion are probably vulnerable to invasion such that above some threshold level of propagule input successful invasion is a virtual certainty. Consequently, this relationship must be nonlinear because by definition it saturates at a probability of one. In our system disturbance appeared to reduce the level of propagule pressure necessary to ensure invasion success. However, even control plots had a high probability of invasion once the level of propagule pressure exceeded that produced by an average adult *S. muticum*. Unfortunately, the limited number of treatment levels in our experiment constrains our ability to resolve the details of this relationship. Nevertheless, in the control treatment there was some evidence of a threshold level of propagule pressure below which invasion was very unlikely to occur (Fig. 3).

Our model reflects what we believe to be the most important factors limiting invasion success (propagule-limitation and competition for space) but other factors we did not include in the model, such as stochastic mortality, density-dependent mortality of adults, competition with native species for resources besides space (e.g. light, Britton-Simmons 2006) and abiotic conditions, could constrain *S. muticum*'s distribution and abundance in the field. Empirical studies have demonstrated the importance of biotic resistance in regulating invasions (see reviews by Levine & D'Antonio 1999; Levine *et al.* 2004) and the community that *S. muticum* is invading is no exception (Britton-Simmons 2006). However, some authors have suggested that propagule pressure has the potential to overcome biotic resistance (D'Antonio *et al.* 2001; Lockwood *et al.* 2005). Levine (2000) found that seed

supply overpowered biotic resistance that was generated by plant communities at small spatial scales (18 cm × 18 cm). A more recent terrestrial experiment also reported that propagule pressure was the primary determinant of invasion success, overwhelming the effects of other factors, such as disturbance and resident diversity, which were concurrently manipulated (Von Holle & Simberloff 2005). However, 'propagules' in that study were seedlings transplanted into experimental plots and seedlings may not be regulated by the same factors as seeds, which are the life stage responsible for invasion spread in natural systems. Nevertheless, if propagule pressure can indeed overcome those factors that were not included in our model then one might ask why *S. muticum* has not completely taken over the shallow subtidal zone in this system, as our model predicts under most disturbance regimes. Interestingly, whether *S. muticum* is indeed in the process of doing so is not entirely clear. There are very few areas in the San Juan region where *S. muticum* is completely absent at the appropriate depths (personal observation), yet at many sites *S. muticum* is currently at low abundance and it is unclear whether these sites represent incipient invasions or whether something is inhibiting local population growth.

Conclusions

In our system, neither disturbance nor propagule input alone was sufficient to maximize invasion success (i.e. establishment probability and invader population density). Increasing propagule pressure had relatively little effect on total recruitment in control plots (Fig. 1a), though at high levels it ultimately overcame space limitation and ensured successful invasion (Fig. 3). However, even at high levels of propagule input, final *S. muticum* density was low in the absence of disturbance (Fig. 1b). Based on our experimental results alone, we might have predicted strong effects of both molluscs and urchins on the *S. muticum* invasion in the long term. However, the simulation model suggested that these two natural disturbance agents should have different effects on long-term invasion due to differences in the spatial structure of these disturbances. The model results demonstrate that caution should be exercised when extrapolating the results of short-term disturbance experiments over longer time intervals. In this marine community invasion success was co-regulated by propagule pressure and biotic resistance. Our results underscore the importance of considering interactive effects when making predictions about invasion success.

Acknowledgements

Thanks to Ben Pister, Sam Sublett and Jake Gregg for SCUBA assistance in the field. For helpful discussions that improved this work we thank Timothy Wootton, Cathy Pfister, Greg Dwyer, Joy Bergelson, Mathew Leibold, Spencer Hall and Bret Elderd. Yvonne Buckley, Barney Davies and an anonymous referee provided very helpful comments on an earlier version of the manuscript. The director and staff of Friday Harbor Laboratories provided logistical support and access to laboratory and SCUBA facilities. The field research was funded by a grant to Timothy Wootton from The Sea Doc Society at UC Davis, and both authors were supported by a Graduate Assistance in Areas of National Need Training Grant (P200A040070) during the completion of this work.

References

Ahlroth, P., Alatalo, R., Holopainen, A., Kumpulainen, T. & Suhonen, V. (2003) Founder population size and number of source populations enhance colonization success in waterstriders. Oecologia, 137, 617–620.

Bergelson, J., Newman, J.A. & Floresroux, E.M. (1993) Rates of weed spread in spatially heterogeneous environments. Ecology, 74, 999–1011.

Bossenbroek, J.M., Kraft, C.E. & Nekola, J.C. (2001) Prediction of long-distance dispersal using gravity-models: zebra mussel invasion of inland lakes. Ecological Applications, 11, 1778–1788.

Britton-Simmons, K.H. (2004) Direct and indirect effects of the introduced alga *Sargassum muticum* on benthic, subtidal communities of Washington State, USA. Marine Ecology Progress Series, 277, 61–78.

Britton-Simmons, K.H. (2006) Functional group diversity, resource preemption and the genesis of invasion resistance in a community of marine algae. Oikos, 113, 395–401.

Brown, R.L. & Peet, R.K. (2003) Diversity and invasibility of southern Appalachian plant communities. Ecology, 84, 32–39.

Buckley, Y.M., Bolker, B.M. & Rees, M. (2007) Disturbance, invasion, and reinvasion: managing the weed-shaped hole in disturbed ecosystems. Ecology Letters, 10, 809–817.

Burnham, K.P & Anderson, D.R. (2002) Model Selection and Inference: A Practical Information–Theoretic Approach. Springer Publishing, New York, NY.

Cassey, P, Blackburn, T.M., Duncan, R.P & Lockwood, J.L. (2005) Lessons from the establishment of exotic species: a meta-analytical case study using birds. Journal of Animal Ecology, 74, 250–258.

Clark, G.F & Johnston, E.L. (2005) Manipulating larval supply in the field: a controlled study of marine invasibility. Marine Ecology Progress Series, 298, 9–19.

D'Antonio, C.M., Dudley, T.L. & Mack, M. (1999) Disturbance and biological invasions: direct effects and feedbacks. Ecosystems of the World 16: Ecosystems of Disturbed Ground (ed. L.R. Walker), pp. 413–452. Elsevier.

D'Antonio, C.M., Levine, J. & Thomsen, V. (2001) Ecosytem resistance and the role of propagule supply: a California perspective. Journal of Mediterranean Ecology, 2, 233–245.

Davis, M.A., Grime, J.P. & Thomson, K. (2000) Fluctuating resources in plant communities: a general theory of invasibility. Journal of Ecology, 88, 528–534.

De Wreede, R.E. (1983) *Sargassum muticum* (Fucales, Phaeophyta): regrowth and interaction with *Rhodomela larix* (Ceramiales, Rhodophyta). Phycologia, 22 (2), 153–160.

Deysher, L. & Norton, T.A. (1982) Dispersal and colonization in *Sargassum muticum* (Yendo) Fensholt. Journal of Experimental Marine Biology and Ecology, 56 (2–3), 179–195.

Drake, J.M., Baggenstos, P & Lodge, D.M. (2005) Propagule pressure and persistence in experimental populations. Biology Letters, 1, 480–483.

Duggins, D.O. (1980) Kelp beds and sea otters: an experimental approach. Ecology, 61, 447–453.

Duggins, D.O. (1983) Starfish predation and the creation of mosaic patterns in a kelp-dominated community. Ecology, 64, 1610–1619.

Duggins, D.O., Eckman, J.E., Siddon, C.E. & Klinger, T. (2003) Population, morphometric and biomechanical studies of three understory kelps along a hydrodynamic gradient. Marine Ecology Progress Series, 265, 57–76.

Foreman, R.E. (1977) Benthic community modification and recovery following intensive grazing by *Strongylocentrotus droebachiensis*. Helgoländer wiss. Meeresunters, 30, 468–484.

Gentle, C.B. & Duggin, J.A. (1997) *Lantana camara* L. invasions in dry rainforest-open forest ecotones: the role of disturbances associated with fire and cattle grazing. Australian Journal of Ecology, 22, 298–306.

Grevstad, F.S. (1999) Experimental invasions using biological control introductions: the influence of release size on the chance of population establishment. Biological Invasions, 1, 313–323.

Griffith, B., Scott, J.M., Carpenter, J.W. & Reed, C. (1989) Translocation as a species conservation tool: status and strategy. Science, 245 (4917), 477–480.

Higgins, S.I. & Richardson, D.M. (1998) Pine invasions in the southern hemisphere: modelling interactions between organism, environment and disturbance. Plant Ecology, 135, 79–93.

Hill, S.J., Tung, P.J. & Leishman, M.R. (2005) Relationships between anthropogenic disturbance, soil properties and plant invasion in endangered Cumberland plain woodland, Australia. Austral Ecology, 30, 775–788.

Hobbs, R.J. & Huenneke, L.F. (1992) Disturbance, diversity, and invasion: implications for conservation. Conservation Biology, 6, 324–337.

Kotanen, P.M. (1997) Effects of experimental soil disturbance on revegetation by natives and exotics on coastal Californian meadows. Journal of Applied Ecology, 34, 631–644.

Larson, D.L. (2003) Native weeds and exotic plants: relationships to disturbance in mixed-grass prairie. Plant Ecology, 169, 317–333.

Levin, P.S., Coyer, J.A., Petrik, R. & Good, T.P. (2002) Community-wide effects of nonindigenous species on temperate rocky reefs. Ecology, 83, 3182–3193.

Levine, J.M. (2000) Species diversity and biological invasions: relating local process to community pattern. Science, 288, 852–854

Levine, J., Adler, P. & Yelenik, S. (2004) A meta-analysis of biotic resistance to exotic plant invasions. Ecology Letters, 7, 975–989.

Levine, J.M. & D'Antonio, C.M. (1999) Elton revisited: a review of evidence linking diversity and invasibility. Oikos, 87, 15–26.

Lockwood, J.L., Cassey, P. & Blackburn, T. (2005) The role of propagule pressure in explaining species invasions. Trends in Ecology and Evolution, 20, 223–228.

Lonsdale, W.M. (1999) Global patterns of plant invasions and the concept of invasibility. Ecology, 80, 1522–1536.

MacDonald, I., Loope, L., Usher, M. & Hamann, O. (1989) Wildlife conservation and the invasion of nature reserves by introduced species: a global perspective. Biological Invasions: A global perspective (eds J.A. Drake, H.A. Mooney, F. di Castri, R.H. Groves, F.J. Kruger, M. Rejmanek & M. Williamson), pp. 215–255. John Wiley & Sons, Chichester, UK.

Mack, M., Simberloff, D., Lonsdale, W., Evans, H., Clout, M. & Bazzaz, F. (2000) Biotic invasions: causes, epidemiology, global consequences, and control. Ecological Applications, 10, 689–710.

Melbourne, B.A., Cornell, H.V., Davies, K.F., Dugaw, C.J., Elmendorf, S., Freestone, A.L., Hall, R.J., Harrison, S., Hastings, A., Holland, M., Holyoak, M., Lambrinos, J., Moore, K. & Yokomizo, H. (2007) Invasion in a heterogeneous world: resistance, coexistence or hostile takeover? Ecology Letters, 10, 77–94.

Naeem, S., Knops, J., Tilman, D., Howe, K., Kennedy, T. & Gale, S. (2000) Plant diversity increases resistance to invasion in the absence of covarying extrinsic factors. Oikos, 91, 97–108.

Nearshore Habitat Program (2001) The Washington State ShoreZone Inventory. Washington State Department of Natural Resources, Olympia, WA.

Norton, T.A. (1977) Ecological experiments with *Sargassum muticum*. Journal of the Marine Biology Association of the United Kingdom, 57, 33–43.

Norton, T.A. & Deysher, L.E. (1988) The reproductive ecology of *Sargassum muticum* at different latitudes. Reproduction, Genetics and Distributions of Marine Organisms: 23 rd European Marine Biology Symposium, School of Biological Sciences, University of Wales, Swansea (eds J.S. Ryland & P.A. Tyler), pp. 147–152. Olsen & Olsen, Fredensborg, Denmark.

Paine, R.T. & Vadas, R.L. (1969) The effects of grazing by sea urchins, *Strongylocentrotus* spp., on benthic algal populations. Limnology and Oceanography, 14, 710–719.

Panetta, F.D. & Randall, R.P. (1994) An assessment of the colonizing ability of *Emex australis*. Australian Journal of Ecology, 19, 76–82.

Parker, I.M. (2001) Safe site and seed limitation in *Cystisus scoparius* (Scotch Broom): Invasibility, disturbance, and the role of cryptograms in a glacial outwash prairie. Biological Invasions, 3, 323–332.

Pimentel, D., Zuniga, R. & Morrison, D. (2005) Update on the environmental and economic costs associated with alien-invasive species in the United States. Ecological Economics, 52, 273–288.

Prieur-Richard, A. & Lavorel, S. (2000) Invasions: the perspective of diverse plant communities. Austral Ecology, 25, 1–7.

Richardson, D.M. & Bond, W.J. (1991) Determinants of plant distribution: evidence from pine invasions. American Naturalist, 137, 639–668.

Ruiz, G.M. & Carlton, J.T. (2003) Invasion vectors: a conceptual framework for management. Invasive Species: Vectors and Management Strategies (eds G.M. Ruiz & J.T. Carlton), pp. 459–504. Island Press.

Sanchez, I. & Fernandez, C. (2006) Resource availability and invasibility in an intertidal macroalgal assemblage. Marine Ecology Progress Series, 313, 85–94.

Scagel, R.F. (1956) Introduction of a Japanese alga, *Sargassum muticum*, into the Northeast Pacific. Washington Department of Fisheries, Fisheries Research Papers, 1, 1–10.

Simberloff, D. (1989) Which insect introductions succeed and which fail? Biological Invasions: A global perspective, SCOPE37 (eds J.A. Drake, H.A. Mooney, F. di Castri, R.H. Groves, F.J. Kruger, M. Rejmanek and M. Williamson), pp. 61–75. John Wiley & Sons, Chichester, UK.

Sjøtun, K., Eggereide, S.F. & H0isaeter, T. (2007) Grazer-controlled recruitment of the introduced *Sargassum muticum* (Phaeophyceae, Fucales) in northern Europe. Marine Ecology Progress Series, 342, 127–138.

Stachowicz, J., Whitlatch, R. & Osman, R. (1999) Species diversity and invasion resistance in a marine ecosystem. Science, 286, 1577–1579.

Thomsen, M.A., D'Antonio, C.M., Suttle, K.B. & Sousa, W.P. (2006) Ecological resistance, seed density and their interactions determine patterns of invasion in a California grassland. Ecology Letters, 9, 160–170.

Travis, J.M.J., Hammershoj, M. & Stephenson, C. (2005) Adaptation and propagule pressure determine invasion dynamics: insights from a spatially explicit model for sexually reproducing species. Evolutionary Ecology Research, 7, 37–51.

Vadas, R.L. (1968) The ecology of Agarum and the kelp bed community. PhD Dissertation, University of Washington.

Valentine, J.P. & Johnson, C.R. (2003) Establishment of the introduced kelp *Undaria pinnatifida*. Tasmania depends on disturbance to native algal assemblages. Journal of Experimental Marine Biology and Ecology, 265, 63–90.

Valentine, J.P. & Johnson, C.R. (2005) Persistence of the exotic kelp *Undaria pinnatifida* does not depend on sea urchin grazing. Marine Ecology Progress Series, 285, 43–55.

Veltman, C.J., Nee, S. & Crawley, M.J. (1996) Correlates of introduction success in exotic New Zealand birds. American Naturalist, 147, 542–557.

Von Holle, B. & Simberloff, D. (2005) Ecological resistance to biological invasion overwhelmed by propagule pressure. Ecology, 86, 3212–3218.

Williamson, M. (1989) Mathematical models of invasio. Biological Invasions: A global perspective (ed. by J.A. Drake, H.A. Mooney, F. di Castri, R.H. Groves, F.J. Kruger, M. Rejmanek and M. Williamson), pp. 329–350. John Wiley & Sons, Chichester, UK.

Williamson, M. (1996) Biological Invasions. Chapman & Hall, London, UK.

Williamson, J. & Harrison, S. (2002) Biotic and abiotic limits to the spread of exotic revegetation species. Ecological Applications, 12, 40–51.

Received 12 June 2007; accepted 1 October 2007
Handling Editor: Jonathan Newman

Supplementary material

The following supplementary material is available for this article:

Appendix S1. Detailed diet information for benthic, subtidal mollusc species.

Appendix S2. Model parameter values and functions.

Appendix S3. Models for disturbance.

This material is available as part of the online article from:
http://www.blackwell-synergy.com/doi/abs/10.1111/j.1365-2745.2007.01319.x
(This link will take you to the article abstract).

Please note: Blackwell Publishing is not responsible for the content or functionality of any supplementary materials supplied by the authors. Any queries (other than missing material) should be directed to the corresponding author for the article.

© 2007 The Authors. Journal compilation © 2007 British Ecological Society, *Journal of Ecology*, 96, 68–77

CHAPTER 21

PEA3: Ganci et al. (2012)

Reprinted from Gaetana Ganci, Annamaria Vicari, Annalisa Cappello, Ciro Del Negro.
An emergent strategy for volcano hazard assessment: From thermal satellite monitoring
to lava flow modelling. *Remote Sensing of Environment*, Volume 119, 2012, Pages 197–207,
ISSN 0034-4257, https://doi.org/10.1016/j.rse.2011.12.021.

Writing Scientific Research Articles: Strategy and Steps, Third Edition.
Margaret Cargill and Patrick O'Connor.
© 2021 John Wiley & Sons Ltd. Published 2021 by John Wiley & Sons Ltd.

Remote Sensing of Environment 119 (2012) 197–207

Contents lists available at SciVerse ScienceDirect

Remote Sensing of Environment

Journal homepage: www.elsevier.com/locate/rse

An emergent strategy for volcano hazard assessment: From thermal satellite monitoring to lava flow modeling

Gaetana Ganci[a,*], Annamaria Vicari[a], Annalisa Cappello[a,b] and Ciro Del Negro[a]

[a] Istituto Nazionale di Geofisica e Vulcanologia, Sezione di Catania, Osservatorio Etneo, Italy
[b] Dipartimento di Matematica e Informatica, Università di Catania, Italy

ARTICLE INFO

Article history:
Received 8 September 2011
Received in revised form 21 December 2011
Accepted 23 December 2011
Available online 26 January 2012

Keywords:
Etna volcano
Infrared remote sensing
Numerical simulation
GIS
Lava hazard assessment

ABSTRACT

Spaceborne remote sensing techniques and numerical simulations have been comined in a web-GIS framework (LAV@HAZARD) to evaluate lava flow hazard in real time. By using the HOTSAT satellite thermal monitoring system to estimate time-varying TADR (tine average discharge rate) and the MAGFLOW physics-based model to simulate lava flow paths, the LAV@Hazard platform allows timely definition of parameters and maps essential for hazard assessment, including the propagation time of lava flows and the maximum run-out distance. We used LAV@HAZARD during the 20082209 lava flow-forming eruption at Mt Etna (Sicily, Italy). We measured the temporal variation in thermal emission (up to four times per hour) during the entire duation of the eruption using SEVIRI and MODIS data. The time-series of radiative power allowed us to identify six diverse thermal phases each related to different dynamic volcanic processes and associated with different TADRs and lave flow emplacment conditions. Satellite-derived estimates of lave discharge rates were computed and intergrated for the whole period of the eruption (almost 14 months), showing that a lava volume of between 32 and 61 million cubic meters was erupted of which about 2/3 was emplaced during the first 4 months. These time-varying discharge rates were then used to drive MAGFLOW simulations to chart the spread of lava as a function of tiime. TADRs were sufficiently low (<20 m³/s) that no lava flows were capable of flowing any great distance so that they did not pose a hazard to vulnerable (agricultural and urban) areas on the flanks of Etna.

© 2012 Elsevier Inc. All rights reserved.

1 Introduction

Etna volcano, Italy, is characterized by persistent activity, consisting of degassing and explosive phenomena at its summit craters, as well as frequent flank eruptions. All eruption typologies can give rise to lava flows, which are the greatest hazard presented by the volcano to inhabited and cultivated areas (Behncke et al., 2005). The frequent eruptions of Mt Etna and hazard they pose represent an excellent opportunity to apply both spaceborne remote sensing techniques for thermal volcano monitoring, and numerical simulations for predicting the area most likely to be inundated by lava during a volcanic eruption. Indeed, Etna has witnessed many tests involving the application of space-based remote sensing data to detect, measure and track the thermal expression of volcanic effusive phenomena (e.g. Ganci et al., 2011b; Harris et al., 1998; Wright et al., 2004), as well as a number of lava flow emplacement models (e.g. Crisci et al., 1986; Harris & Rowland, 2001; Vicari et al., 2007). These efforts have allowed key at-risk areas to be rapidly and appropriately identified (Ganci et al., 2011b; Wright et al., 2008).

Over the last 25 years, satellite measurements in the thermal infrared have proved well suited to detection of volcanic thermal phenomena (e.g. Francis & Rothery, 1987; Harris et al., 1995) and to map the total thermal flux from active lava flows (e.g. Flynn et al., 1994; Harris et al., 1998). High spatial resolution data collected by Landsat and ASTER have been employed for the thermal analysis of active lava flows (Hirn et al., 2007; Oppenheimer, 1991), lava domes (Carter & Ramsey, 2010; Kaneko et al., 2002; Oppenheimer et al., 1993), lava lakes (Harris et al., 1999; Wright et al., 1999), and fumarole fields (Harris & Stevenson, 1997; Pieri & Abrams, 2005). Lower spatial, but higher temporal, resolution sensors, such the Advanced Very High Resolution Radiometer (AVHRR) and the MODerate Resolution Imaging Spectro-radiometer (MODIS), have also been widely used for infrared remote sensing of volcanic thermal features, as has GOES (e.g. Rose & Mayberry, 2000). The high temporal resolution (15 min) offered by the Spinning Enhanced Visible and Infrared Imager (SEVIRI), already employed for the thermal monitoring of effusive volcanoes in Europe and Africa (Hirn et al., 2008), has recently been exploited to estimate lava discharge rates for eruptive events of short duration (a few hours) at Mt Etna (Bonaccorso et al., 2011a, 2011b). These data have also proved capable of detecting, measuring and tracking volcanic thermal phenomena, despite the fact that the volcanic thermal phenomena are usually much smaller than the

* Corresponding author at: Istituto Nazionale di Geofisica e Vulcanologia (INGV), Piazza Roma 2, 95123 Catania, Italy. Tel.: +39 095 7165800; fax +39 095 435801.
E-mail address: gaetana.ganci@ct.ingv.it (G. Ganci).

0034-4257/$ – see front matter © 2012 Elsevier Inc. All rights reserved.
doi:10.1016/j.rse.2011.12.021

nominal pixel sizes of 1 to 4 km (Ganci et al., 2011a, 2011b; Vicari et al., 2011b). Great advances have also been made in understanding the physical processes that control lava flow emplacement, resulting in the development of a range of tools allowing the assessment of lava flow hazard. Many methods have been developed to simulate and predict lava flow paths and run-out distance based on various simplifications of the governing physical equations and on analytical and empirical modeling (e.g.: Hulme, 1974; Crisci et al., 1986; Young & Wadge, 1990; Miyamoto & Sasaki, 1997; Harris & Rowland, 2001; Costa & Macedonio, 2005; Del Negro et al., 2005). In its most simple form, such forecasting may involve the application of volcano-specific empirical length/effusion rate relationships to estimate flow length (e.g. Calvari & Pinkerton, 1998). At their most complex, such predictions may involve iteratively solving a system of equations that characterize the effects that cooling-induced changes in rheology have on the ability of lava to flow downhill, while taking into account spatial variations in slope determined from a digital elevation model (e.g. Del Negro et al., 2008).

On Etna, the MAGFLOW Cellular Automata model has successfully been used to reproduce lava flow paths during the 2001, 2004 and 2006 effusive eruptions (Del Negro et al., 2008; Herault et al., 2009; Vicari et al., 2007). More recently, it has been applied to consider the impact of hypothetical protective barrier placement on lava flow diversion (Scifoni et al., 2010) and has been used for the production of a hazard map for lava flow invasion on Etna (Cappello et al., 2011a, 2011b). MAGFLOW is based on a physical model for the thermal and rheological evolution of the flowing lava. To determine how far lava will flow, MAGFLOW requires constraint of many parameters. However, after chemical composition, the instantaneous lava output at the vent is the principal parameter controlling the final dimensions of a lava flow (e.g. Harris & Rowland, 2009; Pinkerton & Wilson, 1994; Walker, 1973). As such, any simulation technique that aims to provide reliable lava flow hazard assessments should incorporate temporal changes in discharge rate into its predictions in a timely manner.

Satellite remote sensing provides a means to estimate this important parameter in real-time during an eruption (e.g. Harris et al., 1997; Wright et al., 2001). A few attempts to use satellite-derived discharge rates to drive numerical simulations have already been made (e.g. Herault et al., 2009; Vicari et al., 2009; Wright et al., 2008). These modeling efforts all used infrared satellite data acquired by AVHRR and/or MODIS to estimate the time-averaged discharge rate (TADR) following the methodology of Harris et al. (1997). Recently, we developed the HOTSAT multiplatform system for satellite infrared data analysis, which is capable of managing multispectral data from different sensors as MODIS and SEVIRI to detect volcanic hot spots and output their associated radiative power (Ganci et al., 2011b). Satellite-derived output for lava flow modeling purposes has already been tested for the 12–13 January 2011 paroxysmal episode at Mt Etna by Vicari et al., 2011b. Here we present a new web-based GIS framework, named LAV@HAZARD, which integrates the HOTSAT system for satellite-derived discharge rate estimates with the MAGFLOW model to simulate lava flow paths. As a result, LAV@HAZARD now represents the central part of an operational monitoring system that allows us to map the probable evolution of lava flow-fields while the eruption is ongoing. Here we describe and demonstrate the operation of this LAV@HAZARD using a retrospective analysis of Etna's 2008–2009 flank eruption. This eruption was exceptionally well documented by a variety of monitoring techniques maintained by INGV-CT including spaceborne thermal infrared measurements. Within this operational role, HOTSAT is first used to identify thermal anomalies due to active lava flows and to compute the TADR. This is then used to drive MAGFLOW simulations, allowing us to effectively simulate the advance rate and maximum length for the active lava flows. Because satellite-derived TADRs can be obtained in real times and simulations spanning several days of eruption can be calculated in a few minutes, such a combined approach has the potential to provide timely predic-

tions of the areas likely to be inundated with lava, which can be updated in response to changes of the eruption conditions as detected by the current image conditions. If SEVIRI data are used, simulations can be updated every 15 min. Our results thus demonstrate how LAV@HAZARD can be exploited to produce realistic lava flow hazard scenarios and for helping local authorities in making decisions during a volcanic eruption.

2 HOTSAT satellite monitoring system

HOTSAT is an automated system that ingests infrared data acquired by MODIS and SEVIRI data. The decision to employ both SEVIRI (Govaerts et al., 2001) and MODIS (Running et al., 1994) sensors is due to the advantages furnished by their different resolutions. In particular, the high temporal resolution offered by SEVIRI enables almost continuous monitoring (i.e. up to four times per hour) of volcanic thermal features allowing us to measure short and rapidly evolving eruptive phenomena. On the other hand, the higher spatial resolution (1 km, as opposed to 3 km, pixels), the good spectral resolution and the high signal to noise ratio of MODIS permit to detect less intense thermal anomalies and to locate them with more detail. Moreover, the "fire" channel of MODIS (channel 21, 3.9 µm) is designed with a higher saturation temperature of about 500 K (with respect to SEVIRI channel 4, which is also located at 3.9 µm and saturated at about 335 K).

2.1 Hot spot detection

The automatic detection in satellite images of hot spots (thermal anomalies that possibly relate to dynamic volcanic processes) is a nontrivial question, since an appropriate threshold radiance value must be chosen to reveal the pixels containing thermal anomalies. Traditional algorithms rely on simple threshold tests, and open questions arise from the right choice of the thresholds based on latitude, season and time of the day. To overcome limits due to a fixed threshold, we applied a contextual algorithm to both MODIS and SEVIRI images. As starting point, we take the contextual approach of Harris et al. (1995), where a non-volcanic area is defined in order to calculate a threshold from within the image. The algorithm computes for each pixel the difference (ΔT_{diff}) between brightness temperature in mid-infrared (MIR: 3.9 µm) and thermal infrared (TIR: 10–12 µm), and the spatial standard deviation (SSD) of ΔT_{diff}. A threshold (SSD_{max}) is then defined as the maximum of $SSD(\Delta T_{diff})$ from the non-volcanic area within the image.

To locate thermal anomalies automatically, we introduced a two-step approach (Ganci et al., 2011b). In a first step, all the pixels belonging to the "volcanic" area of the image are scanned and those having a value of $SSD(\Delta T_{diff})$ greater than SSD_{max} are classified as "potentially" hot. In a second step, all the "potentially" hot pixels and their neighbors are analyzed to assess whether the potential hot spot is valid. The detection is considered valid if at least one of two conditions was met:

$$T_{3.9\,\mu m} - \min\left(T_{3.9\,\mu m}\right) > MaxVar\left(T_{3.9\,\mu m}\right) \tag{1}$$

$$T_{3.9\,\mu m} > mean\left(T_{3.9\,\mu m}\right) + n * std\left(T_{3.9\,\mu m}\right) \tag{2}$$

where $T_{3.9\mu m}$ is the 3.9 µm brightness temperature, and $MaxVar(T_{3.9\mu m})$ and $std(T_{3.9\mu m})$ are respectively the maximum variation and the standard deviation of the 3.9 µm brightness temperature computed in the non-volcanic area of the image. Parameter n controls how much the MIR pixel-integrated temperature deviates from the mean value. Following this procedure, all the computations are based on dynamic thresholds calculated for the image in hand.

2.2. Heat flux and TADR conversion

HOTSAT next calculates the radiative power for all "hot" pixels. To do this, we applied the MIR radiance conversion of Wooster et al. (2003), an approach that allows an estimate of the radiative power from a sub-pixel hot spot using an approximation of Planck's Law. For each hot spot pixel, the radiative power from all hot thermal components is calculated by combining Stefan–Boltzmann law and Planck's Law, obtaining:

$$FRP_{MIR} = \frac{A_{sampl}\,\varepsilon\sigma}{a\varepsilon_{MIR}} L_{MIR,b} \qquad (3)$$

where, $Qp_{ixe}i$ is the radiative power radiated by the thermally anomalous pixel [W], A_{sampl} is the ground sampling (i.e. pixel) area [m²], s is the emissivity, σ is the Stefan-Boltzmann constant [5.67× 10^{-8} J s^{-1} m^{-2} K^{-4}], $L_{MIR,h}$ and ε_{MIR} are the hot pixel spectral radiance and surface spectral emissivity in the appropriate MIR spectral band, and constant a [W m^{-4} sr^{-1} μm^{-1} K^{-1}] is determined from empirical best-fit relationships. We use the value of a as defined by Wooster et al. (2005) for measurements at 3.9 μm. From Eq. (3), the radiative power is proportional to the calibrated radiance associated with the hot part of the pixel computed as the difference between the observed hot spot pixel radiance in the MIR channel and the background radiance, which would have been observed at the same location in the absence of thermal anomalies.

Harris et al. (1997) showed that the total radiative power measured from satellite infrared data can be converted to time-averaged discharge rate (TADR) using:

$$TADR = \frac{Q}{\rho\left(C_p \Delta T + C_L \Delta\varphi\right)} \qquad (4)$$

where Q is the total thermal flux obtained summing up the radiative power computed for each hot spot pixel, p is the lava density, C_p is the specific heat capacity, ΔT is the eruption temperature minus temperature at which flow stops, C_L is the latent heat of crystallization, and ΔΦ is the volume percent of crystals that form while cooling through ΔT. It is worth noting that the conversion from heat flux to volume flux depends on many lava parameters (such as density, specific heat capacity, eruption temperature, etc.) and has to be determined depending on flow conditions (insulation, rheology, slope, crystallinity, etc.) (Harris et al., 2010). As a result, the conversion has generally been reduced to a linear best-fit relation with the form TADR = xQ, where x needs to be set appropriately (Harris & Baloga, 2009; Wright et al., 2001). Given that we cannot fix a single value to characterize this conversion, the most reasonable solution is to use a range of possible values. We thus defined a range of solutions by adopting the extreme values (Table 1) found by Harris et al. (2000, 2007), to be appropriate for calibrating this technique for Etna lavas.

Table 1
Lava parameter values used to convert satellite thermal data to TADR at Mt Etna (following Harris et al., 2000, 2007) and to run MAGFLOW simulations.

Parameter	Description	Value
ρ	Dense rock density	2600 kg m^{-3}
C_p	Specific heat capacity	1150 J kg^{-1} K^{-1}
ϖ	Vesicularity	10–34%
ε	Emissivity	0.9
T_s	Solidification temperature	1173 K
T_e	Extrusion temperature	1360 K
ΔT	Difference between eruption temperature and temperature at which flow is no longer possible	100–200 K
ΔΦ	Crystallization in cooling through ΔT	30–54%
C_L	Latent heat of crystallization	3.5 × 10^5 J kg^{-1}

3 MAGFLOW lava flow simulator

MAGFLOW is a physics-based numerical model for lava flow simulations based on a Cellular Automaton (CA) approach developed at INGV-Catania (Del Negro et al., 2008; Vicari et al., 2007). The MAGFLOW cellular automaton has a two-dimensional structure with cells described by five scalar quantities: ground elevation, lava thickness, heat quantity, temperature, and amount of solidified lava. The system evolution is purely local, with each cell evolving according to its present status and the status of its eight immediate neighbors (i.e. its Moore neighborhood). In this way, the CA can produce extremely complex structures using simple and local rules. The domain (automaton size) needs to be large enough to include the expected maximum extent of the lava flow, and is decomposed into square cells whose width matches the resolution of the Digital Elevation Model (DEM) available for the area. Lava thickness varies according to lava influx between source cells and any neighboring cells. Lava flux between cells is determined according to the height difference in the lava using a steady-state solution for the one-dimensional Navier–Stokes equations for a fluid with Bingham rheology. Conservation of mass is guaranteed both locally and globally (Dragoni et al., 1986).

MAGFLOW uses a thermo-rheological model to estimate the point at which the temperature of a cell drops below a given solidus temperature T_s. This defines the temperature at which lava stops flowing. A corresponding portion of the erupted lava volume remaining in the cell is converted to solid lava. This specifies the total height of the cell but not to the amount of fluid that can move. For our purposes, the thickness of lava remaining in each cell is initially set to zero. Lava flow is then discharged at a certain rate from a cell (or group of cells) corresponding to the vent location in the DEM. The thickness of lava at the vent cell increases by a rate calculated from the volume of lava extruded during each time interval (where the flow rate from vent can change in time). When the thickness at the vent cell reaches a critical level, the lava spreads into neighboring cells. Next, whenever the thickness in any cell exceeds the critical thickness, lava flows into the adjacent cells.

To produce a dynamic picture of probable lava flow paths, MAGFLOW requires constraint of many parameters. These include: (i) a knowledge of the chemical composition of the lava (this places constraints on the eruption temperature of the lava, the relationships between lava temperature and viscosity, and temperature and yield strength, all of which are used to compute the rate at which the lava solidifies), (ii) a digital representation of the topography over which the lava is to be emplaced, (iii) the location of the eruptive vents, and (iv) an estimate of the lava discharge rate. However, it has been well established, and confirmed many times in the literature (see Harris & Rowland, 2009 for review) that, for a given chemical composition, a higher discharge rate allows a volume of lava to spread over a greater area before it solidifies than if fed at a lower discharge rate. As such, simulations that take into account the way in which discharge rate changes during an eruption, and how this influences lava spreading as a function of time, are of special interest, particularly as effusion rates can vary even over relatively short time scales (Harris et al., 2007).

We have just submitted a paper (Bilotta et al., unpublished results) in which the results of a sensitivity analysis of the MAGFLOW Cellular Automaton model are presented. In this companion paper we carry out a sensitivity analysis of the physical and rheological parameters that control the evolution function of the automaton. The results confirm that, for a given composition, discharge rates strongly influence the modeled emplacement. Indeed, to obtain more accurate simulations (all giving the same lava volume), it is better to input into the model a continuous time-varying discharge rate, even if with moderate errors, rather than sparse but accurate measurements. Such records can be created using satellite remote sensing, and can be made available within hours, perhaps minutes, of satellite overpass, especially when using data acquired by low spatial, but high temporal resolution sensors, such as MODIS and SEVIRI.

4 LAV@HAZARD web-GIS framework

We integrated the HOTSAT system and the MAGFLOW model using a web-based Geographical Information System (GIS) framework, named LAV@HAZARD (Fig. 1). In a companion paper, Vicari et al., (2011a) review the technical aspects of the framework. Here we detail the system functionality for tracking an on-going effusive eruption using Etna most recent flank event as an illustrative case-study. By using the satellite thermal monitoring system to estimate time-varying discharge rates and the physics-based model to simulate lava flow paths, our combined approach allows timely definition of parameters essential for hazard monitoring purposes, such as the time of propagation of lava flow fronts, maximum run-out distance, and area of inundation. The choice of the web architecture allows remote control of the whole platform in a rapid and easy way and complete functionality for real-time lava flow hazard assessment. To do this, we exploit the web-based mapping service provided by Google Maps. Due to its widespread use, and due to the fact that it allows a high degree of customization through a number of utilities called Application Programming Interfaces (APIs), Google provides a perfect foundation for our system. The wide array of APIs furnished by Google Maps allows in a very simple way to embed and manipulate maps, and to place over them different information layers.

LAV@HAZARD consists of four modules regarding monitoring and assessment of lava flow hazard at Etna: (i) satellite-derived output by HOTSAT (including time-space evolution of hot spots, radiative power, discharge rate, etc.), (ii) lava flow hazard map visualization, (iii) a database of MAGFLOW simulations of historic lava flows, and (iv) real-time scenario forecasting by MAGFLOW. As part of the satellite module, MODIS and SEVIRI images are automatically analyzed by HOTSAT, which promptly locates the thermally anomalous

pixels. Next, the heat flux from the anomalous pixels is calculated, which is converted to lava discharge rates. Using SEVIRI, these can be calculated, up-date on (and added to) the LAV@HAZARD database up to four times per hour. The satellite-derived discharge rate estimates are then used in the scenario forecasting module as input parameters to MAGFLOW. Because HOTSAT provides minimum and maximum estimates for TADR, two corresponding (end member) lava flow simulations are produced by MAGFLOW. Every time the satellite obtains a new image of Etna volcano, the TADR is updated, and a new pair of simulations is produced.

Although our original implementation of MAGFLOW was intended for serial execution on standard CPUs (Computer Processing Units), the cellular automaton paradigm displays a very high degree of parallelism that makes it suitable for implementation on parallel computing hardware. Thus, in LAV@HAZARD, MAGFLOW has been implemented on Graphic Processing Units (GPUs), these offering very high performances in parallel computing with a total cost of ownership that is significantly inferior to that of traditional computing clusters of equal performance. The porting of MAGFLOW from the original serial code to the parallel computational platforms was accomplished by Bilotta et al. (2011) using CUDA (Compute Unified Device Architecture), a parallel computing architecture provided by NVIDIA Corporation for the deployment of last generations of GPUs as high-performance computing hardware. The benefit of running on GPUs, rather than on CPUs, depends on the extent and duration of the simulated event; while for large, long-running simulations, the GPU can be 70-to-80 times faster, for short-lived eruptions (emplacing lava units of limited areal extent) the increase in speed obtained is between 40-and-50 times. This means that running MAGFLOW on GPUs provides a simulation spanning several days of eruption in a few minutes. In this way, predictions of likely lava flow paths can be promptly produced in a timely fashion and faster than the rate of TADR update (15 min).

Automatic updating of lava flow hazard scenarios and the remote control of HOTSAT and MAGFLOW on the web, make LAV@HAZARD a helpful tool to validate and adjust/refine our output in real time. The

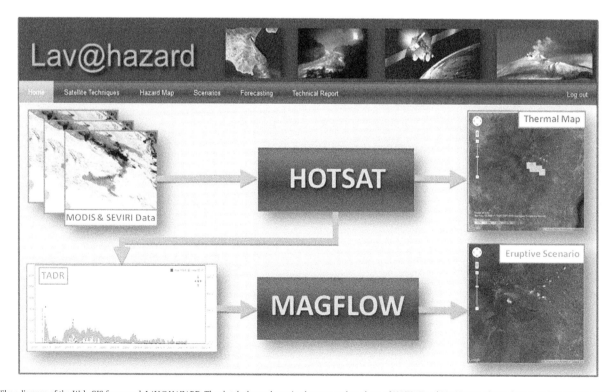

Fig. 1 Flow diagram of the Web-GIS framework LAV@HAZARD. The sketch shows the main elements and products of HOTSAT and MAGFLOW. Thermal maps and TADRs are obtained from MODIS and SEVIRI data. The satellite-derived eruption rates are employed to produce eruptive scenarios.

results of running our web-GIS system during Etna's 2008–2009 eruption are described next.

5 Case study: 2008–2009 Etna eruption

On 10 May 2008 Etna entered an explosive phase, followed by the next effusive eruption 3 days later. Beginning at ~14:00 GMT on 10 May a new vent opened at the eastern base of the South East Crater (SEC) (see Fig. 2). This vent fed intense lava fountaining (where stronger gas jets propelled lava fragments to heights of several hundred meters) characterized by an unusually high rate that in only 4–5 h formed a large lava field. The effusive flank eruption began on 13 May 2008 with the opening of a system of eruptive fissures that propagated SE from the summit craters toward the western wall of the Valle del Bove (VdB). Lava fountains erupted from a fissure extending between 3050 and 2950 m.a.s.l. In 2 h, the eruptive fissure propagated downslope and southeastward, reaching a minimum elevation of 2620 m.a.s.l. (Bonaccorso et al., 2011a). Meanwhile, Strombolian activity migrated from the upper segment of the fissure to its lowest portion. Here, a lava flow erupted at high rates from two main vents (V_1 and V_2 in Fig. 2) and rapidly expanded in the VdB to reach a maximum distance of 6.4 km and extending to 1300 m.a.s.l. in 24 h. Over the following days, field observations highlighted a marked decline in the effusive activity, accompanied by a gradual up-slope migration of the active lava flow fronts and by a decrease in the intensity of the explosive activity from the uppermost portion of the eruptive fissure. Beginning in June and lasting until the end of July, a sudden increase in the effusive activity from the upper vents along the eruptive fissure occurred, causing the lava flow fronts to extend to lower elevations. This period was also associated with an increase in the explosive activity. Over the following month (August), the effusive activity progressively decreased, with only one brief recovery occurring in mid-September 2008. For all the months that followed, the eruptive activity remained relatively constant at low intensity. The eruption ended on 6 July 2009, after almost 14 months of continuous lava effusion.

5.1 Eruption thermal activity

Using the satellite module of LAV@HAZARD, we obtained a detailed chronology of the thermal activity spanning the entire eruption. The analysis of the time series of radiance maps and radiative power, derived from SEVIRI and MODIS images collected from May 2008 to July 2009, allowed detection of diverse thermal phases each of which could be related to different effusive processes:

1. *10 May 2008 (the paroxysmal event).* The sequence of images gathered by SEVIRI on 10 May permitted us to closely follow the opening paroxysmal episode. Volcanic thermal anomalies were detected during the time interval spanning 14:00 to 19:00 GMT. However, the period was characterized by a thick cloud cover and ash emission that led to an underestimation of remotely sensed radiance. Despite the cloudy weather conditions, this fountain was the most powerful of those recorded by SEVIRI sensor between 2007 and 2008 reaching a value, at 15:00 GMT, of ~15 GW (Fig. 3).

2. *13–15 May 2008 (the early effusive phase).* SEVIRI data recorded from 9:15 to 10:30 GMT on 13 May revealed emission of an ash plume (Fig. 4, yellow pixels) that persisted until the first hot spot related to eruptive activity (Fig. 4, red pixels). No eruptive phenomena are observable in the SEVIRI image acquired at 9:15 GMT. The next image acquired at 9:30 GMT detected the beginning of an ash plume moving NE. This was associated with a lava fountain from the northern part of the eruptive fissure. The first hot spot, due to renewed lava output, was detected at 10:30 GMT. Images from MODIS were also analyzed showing a first hot spot at 9:50 GMT (Fig. 5a). High temporal resolution of SEVIRI enabled a precise timing of the early phase of the eruption, but information from MODIS revealed the point of most intense activity, registering a maximum value of radiative power of about 16 GW on 15 May at 00:15 GMT (Figs. 5-6).

3. *16 May–7 June 2008 (the waning phase).* From 16 May, a decline in the SEViRi-recorded radiative power was matched by a change in the structure of the thermal anomaly, with the eastern portion of

Fig. 2 Sketch map of the lava flow field of the 2008-2009 eruption at Mt Etna (see inset). The most distant lava fronts stagnated about 3 km from the nearest village, Milo.

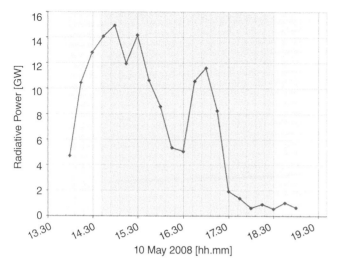

Fig. 3 Total radiative power recorded during the paroxysm occurred at Etna on 10 May 2008. The whole period was characterized by a thick cloud layer (yellow bar) and ash emission (orange bar).

4 *8 June–27 July 2008 (the rising phase).* An increase in radiative power was recorded between 8 June and 27 July, reaching maximum values of about 10 GW on 21 June at 11:35 GMT by MODIS and 6 GW on 13 July at 08:15 GMT by SEVIRI (Fig. 6). This phase was characterized by high but scattered values of radiative power (such as the sudden decrease on 30 June); probably due to discontinuous ash emission during the Strombolian activity, which led to highly variable viewing conditions.

5 *28 July–13 September 2008 (the falling phase).* A progressive decrease in radiative power was observed after 28 July, declining to a minimum value of less than 1 GW on 8 September (Fig. 6). The decreasing trend in the radiative power was interrupted on 11 September, when a peak value of 4 GW was recorded at 08:45 GMT by SEVIRI. This recovery of radiative power was accompanied by a weak explosive activity.

6 *14 September 2008–6 July 2009 (the weak phase).* For all of the months that followed, the SEVIRI-derived radiative power plot showed constantly low values, maintaining a value of about 1 GW (Fig. 6). The last hot spot detected by SEVIRI occurred on 30 May 2009 at 05:00 GMT, with MODIS registering thermal anomalies until 26 June 2009 at 09:45 GMT.

the anomaly becoming less intense. The decrease in radiative power continued until 7 June, maintaining an average daily value of about 1-2 GW (Fig. 6). This is consistent with the low intensity of volcanic activity revealed by field observations during the same period.

Lava discharge rate estimates were computed (up to four times per hour) for the whole period of thermal emission from 13 May 2008 to 26 June 2009, converting the total radiative power measured from infrared satellite data (Fig. 7). We obtained minimum and maximum estimates for TADR by taking into account in the variability range of each lava parameter (Table 1) the largest and smallest

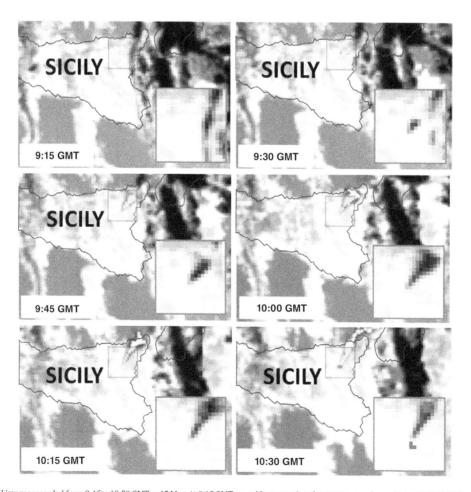

Fig. 4 A sequence of SEVIRI images recorded from 9:15 to 10:30 GMT on 13 May. At 9:15 GMT, no evident eruptive phenomena are observable. At 9:30 GMT the image shows the beginning of an ash plume (yellow pixels) moving toward NE, which was associated to the lava fountain from the northern part of the eruptive fissure. The first hot spot, due to the increased lava output and thermal anomaly, was detected by HOTSAT at 10:30 GMT.

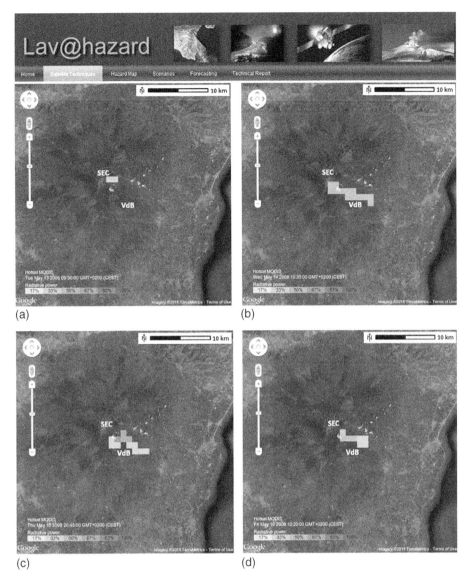

Fig. 5 Hotspot detected during 13–16 May 2008 by MODIS data. Four instants are reported on Google Map: a) 13 May at 09:50 GMT, b) 14 May at 10:35 GMT, c) 15 May at 20:45 GMT, d) 16 May at 10:20 GMT. Different colors are associated to different values of radiative power as reported in the bottom legend.

values, respectively. Peak estimates for TADR were reached on 15 May, ranging between 15 and 28 m³ s⁻¹, and 21 June, ranging between 10 and 18 m³ s⁻¹. By integrating minimum and maximum TADR values over the entire eruptive period preceding each successive measurement, we computed two cumulative curves of erupted lava (Fig. 7). Over the entire period of thermal activity (which ceased on 26 June 2009), we estimate a lava volume of between 32 and 61 million cubic meters was erupted. While erupted volumes by 8 June 2008 were between 3 and 6 million cubic meters, by 13 September they were between 23 and 43 million cubic meters. Showing that about 2/3 of the lava was erupted during the first 4 months of the eruption.

5.2 Lava flow simulations

By using the scenario forecasting module of LAVA@HAZARD, we calculated the lava flow paths using the estimates for TADR derived from thermal satellite data for the first 76 days of eruption, i.e. during the period between 13 May and 27 July 2008. This period of time corresponds to the first three phases of the effusive eruption (phases 2–4) during which time TADR steadily increased to a peak, and by the end of

which the flow-field had attained 100% of its final length. We simulated the evolution of the flow field, step-by-step, over this period using, as our topographic base, a suitable DEM created by up-dating the 2005 2-m resolution DEM with data collected during an aerophotogrammetry flight of 2007 (Neri et al., 2008).

The actual flow areas and two simulated scenarios, obtained by using the minimum and maximum derived TADR, are shown in Fig. 8. The main differences between the simulated and observed lava flow paths are mainly governed by the values of TADR used and topography. In fact, the misfit between the actual and modeled length evolution occurs because the simulated flow is always slower than the actual flow, although the final distribution of lava-inundated areas is almost the same. This is probably due to differences between the maximum value of the satellite-derived lava discharge rate and the real value.

5.3 Discussion

The uncertainty in satellite derived TADR estimates is quite large, up to about 50%, but it is comparable to the error in field-based effusion rate measurements (Calvari et al., 2003; Harris & Baloga, 2009;

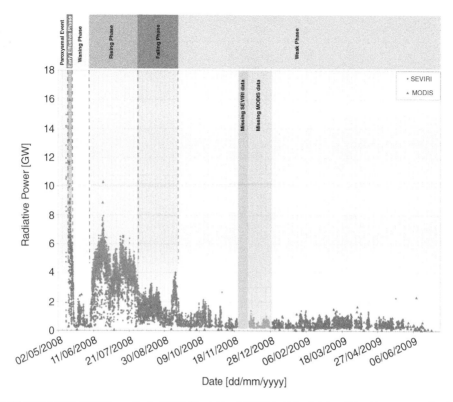

Fig. 6 Radiative power computed during the 2008–2009 Etna eruption by SEVIRI (blue dots) and MODIS (red triangles) data. Colored bars correspond to the six different thermal phases. Gray areas are related to missing data of MODIS (gray) and SEVIRI (dark gray) respectively.

Harris & Neri, 2002; Harris et al., 2007; Sutton et al., 2003). The main uncertainties arise from the lack of constraint on the lava parameters used to convert thermal flux to TADR (Harris & Baloga, 2009). Moreover, the presence of ash strongly affected the satellite-based estimates of lava discharge rate. This led to an underestimation of the satellite-derived final volume, and to a difference in the timing of simulated lava flow emplacement. Moreover, a certain discrepancy between actual and modeled flow areas is to be expected, as changes in the contemporaneous effusion rate take time to be translated into a change in active flow area (that is, the active flow area at any one time is a function of the antecedent effusion rate, rather than the instantaneous effusion rate) (Harris & Baloga, 2009; Wright et al., 2001).

Topographic effects also influence the spatial distribution of the flow. While the spatial overlap between the actual lava flow and the modeled flow driven by the minimum TADR is about 25%, the flow path simulated with the maximum TADR overlaps the actual lava flow by 60% (suggesting that the maximum value is more realistic).

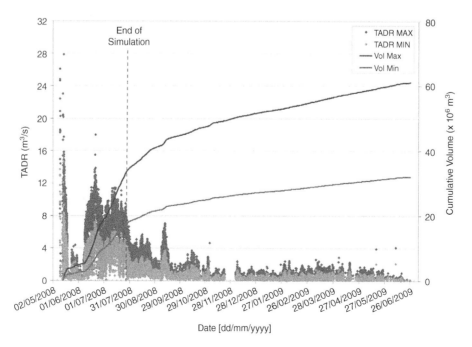

Fig. 7 TADRs and cumulative volumes computed in the period 13 May 2008–26 June 2009 by MODIS and SEVIRI data. Purple and green colors correspond to maximum and minimum values, respectively.

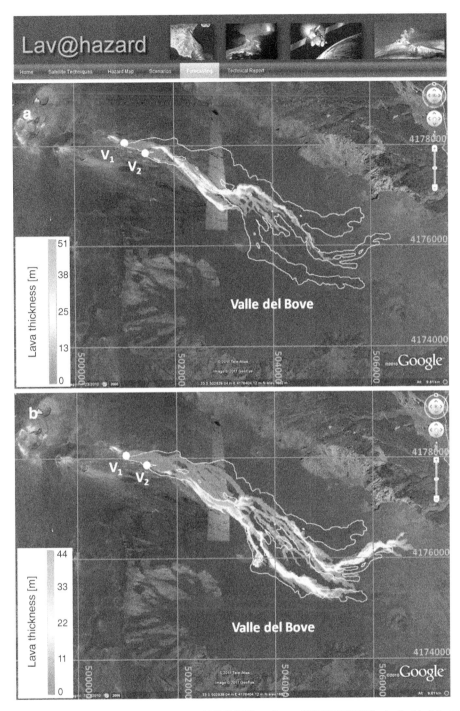

Fig. 8 Lava flow paths simulated by MAGFLOW (V_1 UTM coordinates: 500901E/4177991N, V_2 UTM coordinates: 501263E/4177852N) using the (a) minimum and the (b) maximum TADR. Simulations were performed from 13 May to 27 July 2008 and compared with the actual lava flow field (white contour; Behncke & Calvari, 2008). After this data no significant change appears in the final lava flow path.

The final simulated flow-fields are also always lacking in the northern part mapped for actual flow-field (Fig. 8). These space and time differences were probably due to the DEM, which does not include lava fountains emitted since 2007. These modified the morphology of the upper VdB so that our topography is out-of-date in this northern sector where we experience misfit. However, the zones covered by the simulated lava flow fit well with most of the observed lava flow field.

During the early phase of the eruption, several other simulations were performed to verify, which conditions, especially for the values of TADR, could be considered hazardous for the nearest infrastructures and villages. For all simulated scenarios, no lava flow path extended beyond the VdB area, even if higher values of TADR (increased of 50% of maximum value of satellite-derived TADR) were used as input of MAGFLOW. The most distant lava fronts always stagnated about 3 km from, Milo.

6 Concluding remarks

The LAV@HAZARD web-GIS framework shows great promise as a tool to allow tracking and prediction of effusive eruptions, allowing us to better assess lava flow hazards during a volcanic eruption in real-time. By using satellite-derived discharge rates to drive a lava flow emplacement model, LAV@HAZARD has the capability to forecast the lava flow hazards, allowing the estimation of the inundation area extent, the time required for the flow to reach a particular point, and the resulting morphological changes. We take advantage of the flexibility of the HOTSAT thermal monitoring system to process, in real time, satellite images coming from sensors with different spatial, temporal and spectral resolutions. In particular, HOTSAT was designed to ingest infrared satellite data acquired by the MODIS and SEVIRI sensors to output hot spot location, lava thermal flux and discharge rate. We use LAV@HAZARD to merge this output with the MAGFLOW physics-based model to simulate lava flow paths and to update, in a timely manner, flow simulations. In such a way, any significant changes in lava discharge rate are included in the predictions. MAGFLOW was implemented on the last generation of CUDA-enabled cards allowing us to gain a significant benefit in terms of computational speed thanks to the parallel nature of the hardware. All this useful information has been gathered into the LAV@HAZARD platform which, due to the high degree of interactivity, allows generation of easily readable maps and a fast way to explore alternative scenarios.

We tested LAV@HAZARD in an operational context as a support tool for decision makers using the 2008–2009 lava flow-forming eruption at Mt Etna. By using thermal infrared satellite data with low spatial and high temporal resolution, we continuously measured the temporal changes in thermal emission (at rates of up to four times per hour) for the entire duration of the eruption (also for short-lived phenomena such as the 10 May paroxysm preceding the eruption). The time-series analysis of radiative power, derived from SEVIRI and MODIS, was able to identify six different phases of eruptive activity. Satellite-derived TADR estimates were also computed and integrated for the whole period of the eruption (almost 14 months), showing that about 2/3 of lava volume was erupted during the first 4 months. These time-varying discharge rates were then used in the MAGFLOW model, allowing us to effectively simulate the advance time for active fronts and to produce on-the-fly eruptive scenarios fed by up-to-date TADRs. We found that no lava flows were capable of flowing over distances sufficient to invade vulnerable areas on the flanks of Etna. A comparison with actual, mapped, lava flow-field areas permitted validation of our methodology, predictions and results. Validations confirmed the reliability of LAV@HAZARD and its underlying data source and methodologies, as well as the potential of the whole integrated processing chain, as an effective strategy for real-time monitoring and assessment of volcanic hazard.

Acknowledgment

We are grateful to European Organisation for the Exploitation of Meteorological Satellites (EUMETSAT) for SEVIRI data (www.eumetsat. int) and to National Aeronautics and Space Administration (NASA) for MODIS data (modis.gsfc.nasa.gov). Thanks are due to Maria Teresa Pareschi and Marina Bisson (INGV-Sezione Pisa) for making the Digital Elevation Model of Etna available. This study was performed with the financial support from the V3-LAVA project (DPC-INGV 2007–2009 contract). This manuscript benefited greatly from the encouragement, comments, and suggestions of Andrew Harris, two anonymous reviewers, and the Editor Marvin Bauer.

References

Behncke, B., & Calvari, S. (2008). Etna: 6-km-long lava flow; ash emissions; 13 May 2008 opening of a new eruptive fissure. Bull. Global Volcanism Network, 33(5), 11–15.

Behncke, B., Neri, M., & Nagay, A. (2005). Lava flow hazard at Mount Etna (Italy): New data from a GIS-based study. In M. Manga, & G. Ventura (Eds.), Kinematics and dynamics of lava flows. Spec. Pap. Geol. Soc. Am., 396. (pp. 189–208), doi:10.1130/0-8137-2396-5.189(13).

Bilotta, G., Rustico, E., Hérault, A., Vicari, A., Del Negro, C., & Gallo, G. (2011). Porting and optimizing MAGFLOW on CUDA. Annals of Geophysics, 54, 5, doi:10.4401/ag-5341.

Bilotta, G., Cappello, A., Hérault, A., Vicari, A., Russo, G., & Del Negro, C. (Unpublished results). Sensitivity analysis of the MAGFLOW Cellular Automaton model, Environmental Modelling and Software, submitted for publication, July 2011.

Bonaccorso, A., Bonforte, A., Calvari, S., Del Negro, C., Di Grazia, G., Ganci, G., Neri, M., Vicari, A., & Boschi, E. (2011a). The initial phases of the 2008–2009 Mount Etna eruption: A multidisciplinary approach for hazard assessment. Journal ofGeophysical Research, 116, B03203, doi:10.1029/2010JB007906.

Bonaccorso, A., Caltabiano, T., Currenti, G., Del Negro, C., Gambino, S., Ganci, G., Giammanco, S., Greco, F., Pistorio, A., Salerno, G., Spampinato, S., & Boschi, E. (2011b). Dynamics of a lava fountain revealed by geophysical, geochemical and thermal satellite measurements: The case of the 10 April 2011 Mt Etna eruption. Geophysical Research Letters, 38, doi:10.1029/2011GL049637.

Calvari, S., Neri, M., & Pinkerton, H. (2003). Effusion rate estimations during the 1999 summit eruption on Mount Etna, and growth of two distinct lava flow fields. Journal of Volcanology and Geothermal Research, 119, 107–123, doi:10.1016/S0377-0273(02)00308-6.

Calvari, S., & Pinkerton, H. (1998). Formation of lava tubes and extensive flow field during the 1991–93 eruption of Mount Etna. Journal of Geophysical Research, 103, 27291–27302.

Cappello, A., Vicari, A., & Del Negro, C. (2011a). Retrospective validation of a lava flow hazard map for Mount Etna volcano. Annals of Geophysics, 54, 5, doi:10.4401/ag-5345.

Cappello, A., Vicari, A., & Del Negro, C. (2011b). Assessment and modeling of lava flow hazard on Etna volcano. Bollettino di Geofisica Teorica e Applicata, 52, 299–308, doi:10.4430/bgta0003 n.2.

Carter, A., & Ramsey, M. (2010). Long-term volcanic activity at Shiveluch Volcano: Nine years of ASTER spaceborne thermal infrared observations. Remote Sensing, 2, 2571–2583, doi:10.3390/rs2112571.

Costa, A., & Macedonio, G. (2005). Numerical simulation of lava flows based on depth-averaged equations. Geophysical Research Letters, 32, L05304, doi:10.1029/2004GL021817.

Crisci, G. M., Di Gregorio, S., Pindaro, O., & Ranieri, G. (1986). Lava flow simulation by a discrete cellular model: first implementation. International Journal of Modelling and Simulation, 6, 137–140.

Del Negro, C., Fortuna, L., Herault, A., & Vicari, A. (2008). Simulations of the 2004 lava flow at Etna volcano by the MAGFLOW cellular automata model. Bulletin of Volcanology, doi:10.1007/s00445-007-0168-8.

Del Negro, C., Fortuna, L., & Vicari, A. (2005). Modelling lava flows by Cellular Nonlinear Networks (CNN): Preliminary results. Nonlinear Processes in Geophysics, 12, 505–513.

Dragoni, M., Bonafede, M., & Boschi, E. (1986). Downslope flow models of a Bingham liquid: Implications for lava flows. Journal of Volcanology and Geothermal Research, 30, 305–325.

Flynn, L. P., Mouginis-Mark, P. J., & Horton, K. A. (1994). Distribution of thermal areas on an active lava flow field: Landsat observations of Kilauea, Hawaii, July 1991. Bulletin of Volcanology, 56, 284–296.

Francis, P. W., & Rothery, D. A. (1987). Using the Landsat Thematic Mapper to detect and monitor active volcanoes: An example from Lascar volcano, Northern Chile. Geology, 15, 614–617.

Ganci, G., Vicari, A., Bonfiglio, S., Gallo, G., & Del Negro, C. (2011a). A texton-based cloud detection algorithm for MSG-SEVIRI multispectral images. Geomatics, Natural Hazards and Risk, 2, 1–12, doi:10.1080/19475705.2011.578263 3.

Ganci, G., Vicari, A., Fortuna, L., & Del Negro, C. (2011b). The HOTSAT volcano monitoring system based on a combined use of SEVIRI and MODIS multispectral data. Annals of Geophysics, 54, doi:10.4401/ag-5338 5.

Govaerts, Y., Arriaga, A., & Schmetz, J. (2001). Operational vicarious calibration of the MSG/SEVIRI solar channels. Advances in Space Research, 28, 21–30.

Harris, A. J. L., & Baloga, S. M. (2009). Lava discharge rates from satellite-measured heat flux. Geophysical Research Letters, 36, L19302, doi:10.1029/2009GL039717.

Harris, A. J. L., Blake, S., Rothery, D., & Stevens, N. (1997). A chronology of the 1991 to 1993 Mount Etna eruption using advanced very high resolution radiometer data: implications for real-time thermal volcano monitoring. Journal of Geophysical Research, 102, 7985–8003.

Harris, A.J. L., Dehn, J., & Calvari, S. (2007). Lava effusion rate definition and measurement: A review. Bulletin of Volcanology, doi:10.1007/s00445-007-0120-y.

Harris, A. J. L., Favalli, M., Steffke, A., Fornaciai, A., & Boschi, E. (2010). A relation between lava discharge rate, thermal insulation, and flow area set using lidar data. Geophysical Research Letters, 37, L20308, doi:10.1029/2010GL044683.

Harris, A.J. L., Flynn, L. P., Keszthelyi, L., Mouginis-Mark, P.J., Rowland, S. K., & Resing, J. A. (1998). Calculation of lava effusion rates from Landsat TM data. Bulletin of Volcanology, 60, 52–71.

Harris, A. J. L., Flynn, L. P., Rothery, D. A., Oppenheimer, C., & Sherman, S. B. (1999). Mass flux measurements at active lava lakes: Implications for magma recycling. Journal of Geophysical Research, 104, 7117–7136.

Harris, A. J. L., Murray, J. B., Aries, S. E., Davies, M. A., Flynn, L. P., Wooster, M. J., Wright, R., & Rothery, D. A. (2000). Effusion rate trends at Etna and Krafla and their implications for eruptive mechanisms. Journal of Volcanology and Geothermal Research, 102, 237–270.

Harris, A.J. L., & Neri, M. (2002). Volumetric observations during paroxysmal eruptions at Mount Etna: Pressurized drainage of a shallow chamber or pulsed supply? Journal of Volcanology and Geothermal Research, 116, 79–95, doi:10.1016/S0377-0273(02)00212-3.

Harris, A., & Rowland, S. (2001). FLOWGO: A kinematic thermorheological model for lava flowing in a channel. Bulletin of Volcanology, 63, 20–44.

Harris, A. J. L., & Rowland, S. K. (2009). Effusion rate controls on lava flow length and the role of heat loss: a review. In T. Thordarson, S. Self, G. Larsen, S. K. Rowland, & A. Hoskuldsson (Eds.), Studies in volcanology: The legacy of George Walker. Special Publications of IAVCEI, 2. (pp. 33–51) London: Geological Society978-1-86239-280-9.

Harris, A. J. L., & Stevenson, D. S. (1997). Thermal observations of degassing open conduits and fumaroles at Stromboli and Vulcano using remotely sensed data. Journal of Volcanology Geothermal Research, 76, 175–198.

Harris, A. J. L., Vaughan, R. A., & Rothery, D. A. (1995). Volcano detection and monitoring using AVHRR data: The Krafla eruption, 1984. International Journal of Remote Sensing, 16, 1001–1020.

Herault, A., Vicari, A., Ciraudo, A., & Del Negro, C. (2009). Forecasting lava flow hazard during the 2006 Etna eruption: Using the Magflow cellular automata model. Computer &Geosciences, 35(5), 1050–1060.

Hirn, B., Di Bartola, C., Laneve, G., Cadau, E., & Ferrucci, F. (2008). SEVIRI onboard Meteosat Second Generation, and the quantitative monitoring of effusive volcanoes in Europe and Africa. Proceedings IGARSS 2008 (pp. 4–11). July 2008.

Hirn, B., Ferrucci, F., & Di Bartola, C. (2007). Near-tactical eruption rate monitoring of Pu'u O'o (Hawaii) 2000–2005 by synergetic merge of payloads ASTER and MODIS. Geoscience and Remote Sensing Symposium, IGARSS 2007. IEEE International, 23–28 July 2007. (pp. 3744–3747), doi:10.1109/IGARSS.2007.4423657.

Hulme, G. (1974). The interpretation of lava flow morphology. Geophysical Journal of the Royal Astronomical Society, 39, 361–383.

Kaneko, T., Wooster, M. J., & Nakada, S. (2002). Exogenous and endogenous growth of the Unzen lava dome examined by satellite infrared image analysis. Journal of Volcanology and Geothermal Research, 116, 151–160.

Miyamoto, H., & Sasaki, S. (1997). Simulating lava flows by an improved cellular automata method. Computers & Geosciences, 23, 283–292.

Neri, M., Mazzarini, F., Tarquini, S., Bisson, M., Isola, I., Behncke, B., & Pareschi, M. T. (2008). The changing face of Mount Etna's summit area documented with Lidar technology. Geophysical Research Letters, 35, L09305, doi:10.1029/2008GL033740.

Oppenheimer, C. (1991). Lava flow cooling estimated from Landsat Thematic Mapper infrared data: the Lonquimay eruption (Chile, 1989). Journal of Geophysical Research, 96, 21865–21878.

Oppenheimer, C., Francis, P. W., Rothery, D. A., Carlton, R. W. T., & Glaze, L. S. (1993). Infrared image analysis of volcanic thermal features: Láscarvolcano, Chile, 1984–1992. Journal of Geophysical Research, 98, 4269–4286.

Pieri, D. C., & Abrams, M. J. (2005). ASTER observations of thermal anomalies preceding the April 2003 eruption of Chikurachki Volcano, Kurile Islands, Russia. Remote Sensing of Environment, 99, 84–94.

Pinkerton, H., & Wilson, L. (1994). Factors controlling the lengths of channel-fed lava flows. Bulletin of Volcanology, 56, 108–120.

Rose, W. I., & Mayberry, G. C. (2000). Use of GOES thermal infrared imagery for eruption scale measurements, Soufrière Hills, Montserrat. Geophysical Research Letters, 27, 3097–3100.

Running, S. W., Justice, C., Salomonson, V., Hall, D., Barker, J., Kaufman, Y., Strahler, A., Huete, A., Muller, J. P., Vanderbilt, V., Wan, Z. M., Teillet, P., & Carneggie, D. (1994). Terrestrial remote sensing science and algorithms planned for EOS/MODIS. International Journal of Remote Sensing, 15, 3587–3620.

Scifoni, S., Coltelli, M., Marsella, M., Proietti, C., Napoleoni, Q., Vicari, A., & Del Negro, C. (2010). Mitigation of lava flow invasion hazard through optimized barrier configuration aided by numerical simulation: The case of the 2001 Etna eruption. Journal of Volcanology and Geothermal Research, 192, 16–26, doi:10.1016/j.jvolgeores.2010.02.002.

Sutton, A. J., Elias, T., & Kauahikaua, J. (2003). Lava-effusion rates for the Pu'u'O'o–Kupaianaha eruption derived from SO2 emissions and very low frequency (VLF) measurements. United States Geological Survey Professional Paper, 1676, 137–148.

Vicari, A., Dilotta, C., Bonfiglio, S., Cappello, A., Ganci, G., Herault, A., Rustico, E., Gallo, G., & Del Negro, C. (2011a). LAV@HAZARD: A web-GIS interface for volcanic hazard assessment. Annals of Geophysics, 54, 5, doi:10.4401/ag-5347.

Vicari, A., Ciraudo, A., Del Negro, C., Herault, A., & Fortuna, L. (2009). Lava flow simulations using discharge rates from thermal infrared satellite imagery during the 2006 Etna eruption. Natural Hazards, 50, 539–550, doi:10.1007/s11069-008-9306-7.

Vicari, A., Ganci, G., Behncke, B., Cappello, A., Neri, M., & Del Negro, C. (2011b). Nearreal-time forecasting of lava flow hazards during the 12–13 January 2011 Etna eruption. Geophysical Research Letters, 38, doi:10.1029/2011GL047545.

Vicari, A., Herault, A., Del Negro, C., Coltelli, M., Marsella, M., & Proietti, C. (2007). Simulations of the 2001 lava flow at Etna volcano by the Magflow Cellular Automata model. Environmental Modelling & Software, 22, 1465–1471.

Walker, G. P. L. (1973). Lengths of lava flows. Philosophical Transactions of the Royal Society of London, 274, 107–118.

Wooster, M. J., Roberts, G., Perry, G. L. W., & Kaufman, Y. J. (2005). Retrieval of biomass combustion rates and totals from fire radiative power observations: FRP derivation and calibration relationships between biomass consumption and fire radiative energy release. Geophysical Research Letters, 110, D24311, doi:10.1029/2005JD006318 1–24.

Wooster, M., Zhukov, B., & Oertel, D. (2003). Fire radiative energy release for quantitative study of biomass burning: derivation from the BIRD experimental satellite and comparison to MODIS fire products. Remote Sensing of Environment, 86, 83–107.

Wright, R., Blake, S., Harris, A., & Rothery, D. (2001). A simple explanation for the space-based calculation of lava eruption rates. Earth and Planetary Science Letters, 192, 223–233.

Wright, R., Flynn, L. P., Garbeil, H., Harris, A. J. L., & Pilger, E. (2004). MODVOLC: near-real-time thermal monitoring of global volcanism. Journal of Volcanology and Geothermal Research, 135, 29–9.

Wright, R., Garbeil, H., & Harris, A. J. L. (2008). Using infrared satellite data to drive a thermorheological/stochastic lava flow emplacement model: A method for near-real-time volcanic hazard assessment. Geophysical Research Letters, 35, L19307, doi:10.1029/2008GL035228.

Wright, R., Rothery, D. A., Blake, S., & Harris, A. J. L. (1999). Simulating the response of the EOS Terra ASTER sensor to high-temperature volcanic targets. Geophysical Research Letters, 26, 1773–1776.

Young, P., & Wadge, G. (1990). FLOWFRONT: simulation of a lava flow. Computers & Geosciences, 16, 1171–1191.

Answer pages

Task 2.1 Article headings and subheadings

Headings and subheadings for Kaiser et al. (2003)

Summary
Keywords
Introduction
Results
  Cloning of GmDmt1;1
  Gene expression
  Protein localisation
  Functional analysis in yeast
Discussion
  GmDmt1;1 can transport ferrous iron
  Specificity of GmDmt1;1
  Localisation and function of GmDmt1;1
  Regulation of GmDmt1;1 expression
  Conclusion
Experimental procedures
  Plant growth
  Isolation of GmDmt1;1
  Northern analysis
  Antibody generation and Western immunoblot analysis
  Symbiosome isolation and nodule membrane purification
  Functional expression in yeast
Acknowledgements
References

Headings and subheadings for Britton-Simmons & Abbott (2008)

Summary
Keywords
Introduction
Methods
  Study system
  The invader
  Field experiment
  Statistical analysis

Writing Scientific Research Articles: Strategy and Steps, Third Edition.
Margaret Cargill and Patrick O'Connor.
© 2021 John Wiley & Sons Ltd. Published 2021 by John Wiley & Sons Ltd.

Model
Results
Discussion
  Simulated urchin/mollusc disturbances
  Propagule pressure and invasion success
Conclusions
Acknowledgements
References
Supplementary material

Headings and subheadings for Ganci et al. (2012)

Abstract
1 Introduction
2 HOTSAT satellite monitoring system
  2.1 Hot spot detection
  2.2 Heat flux and TADR conversion
3 MAGFLOW lava flow simulator
4 LAV@HAZARD web-GIS framework
5 Case study: 2008-2009 Etna eruption
  5.1 Eruption thermal activity
  5.2 Lava flow simulations
  5.3 Discussion
6 Concluding remarks
Acknowledgement
References

Task 2.2 Structure of the PEAs

Kaiser et al. (2003) most closely resembles the AIRDaM diagram. Britton-Simmons & Abbott (2008) most closely resembles the AIMRaD diagram, with a separate Conclusions section added at the end. Ganci et al. (2012) most closely matches the AIBC diagram, with four body subsections and the final C called Concluding remarks.

Task 2.3 Prediction

. . . yielded a total of . . .	**(R)**
The aim of the work described . . .	**(I)**
. . . was used to calculate . . .	**(M)** or **(R)**
There have been few long-term studies of . . .	**(I)**
The vertical distribution of . . . was determined by . . .	**(M)** or **(R)**
This may be explained by . . .	**(D)**
Analysis was carried out using . . .	**(M)**
. . . was highly correlated with . . .	**(R)**

Referee criterion	Likely location/s
1 Does the contribution fall within the scope of the journal?	T, A, I, M, D
2 Is the contribution new and original?	I (also stated in A, but no room to demonstrate it there)
3 Is the contribution of sufficiently high impact for the journal?	All, especially A, I, D
4 Does the manuscript demonstrate an adequate awareness of relevant research?	I and D
5 Are the interpretations and conclusions sound and justified by the data presented?	R compared to D and A
6 Is the organisation of the manuscript acceptable?	All
7 Does the title clearly reflect the contents?	T and all
8 Is the abstract sufficiently informative, especially when read in isolation?	A and all
9 Does the introduction set the manuscript in an international context and show how it builds on previous work?	I and Ref
10 Do the methods and analysis conform to acceptable standards and is sufficient detail presented to comprehend the study?	M and R
11 Are all the figures and tables necessary and the captions adequate and informative?	Figures and tables
12 Is the length of the manuscript appropriate to the content?	All
13 Are the references adequate for the subject and length of the manuscript?	Ref and all
14 Is the quality of the English satisfactory?	All

Task 3.2 Information extracted from titles

Title A: Use of in situ[15] N-labelling to estimate the total below-ground nitrogen of pasture legumes in intact soil-plant systems

Information:

- The paper focuses on a particular method (*in situ*[15] N-labelling) and on results obtained using it.
- The parameter measured was total below-ground nitrogen.
- The measurement site/context was undisturbed systems involving both plants and soil.
- The plants used were pasture legumes.

Questions (many others are possible):

- Why is this method suitable to measure this parameter in this context?
- Did the method provide reliable measurements?
- How was the accuracy of the measurements verified?
- How many legumes were studied and how did the results vary between them?

- What soil types were involved?
- Could this method be used for other plant/soil systems?

Title B: Short- and long-term effects of disturbance and propagule pressure on a biological invasion

Information:

- The paper reports the effects of two factors (disturbance and propagule pressure) on one biological invasion.
- Results are reported over two time frames: short-term and long-term.
- The focus of the paper is on generalisations from the findings that apply to biological invasion in general (because no details are given in the title about the specific organisms or sites involved in this particular invasion).

Questions (many others are possible):

- What organisms and locations were involved in the invasion studied?
- What is the meaning of propagule pressure in this context?
- How are "short-term" and "long-term" defined in this paper?
- How do the specific results for this invasion provide evidence for the study of biological invasion in general?

Title C: The soybean NRAMP homologue, GmDMT1, is a symbiotic divalent metal transporter capable of ferrous iron transport

Information:

- The paper reports the function (ability to transport divalent metals) of a newly identified entity: an NRAMP homologue found in soybeans.
- The work reported in the paper shows that the homologue can transport one particular type of iron (ferrous iron).
- The transport process is related to the symbiosis occurring in soybeans.

Questions (many others are possible):

- Why is the transport of ferrous iron significant in soybeans?
- How does the transport of divalent metals relate to the symbiosis?
- How was the function of this entity established?
- How does this finding contribute to the broader study of transporters?

Title D: An emergent strategy for volcano hazard assessment: from thermal satellite monitoring to lava flow modeling

Information:

- The paper reports a method under development for evaluating the dangers presented by volcanoes.
- The work reported in the paper begins with thermal data obtained via satellite monitoring and presents a method for modelling the flow of lava.

Questions (many others are possible):

- What are the advantages of thermal satellite data for this type of application?
- Has the model been tested empirically or is this a theoretical paper?
- What aspects of the strategy are still to be developed?

Improvements in Figure AP.1 are as follows:

- Removal of error bars and replacement with LSD bar decreases clutter, allows comparison of significant differences between treatments, and allows the y axis to be expanded with a lower maximum (i.e. greater spread between the lines). More detail about the significance level of difference is also provided in the figure legend. The removal of the figure border reduces clutter in this line graph.
- The main comparison between chemical fertilizer and chemical fertilizer plus wheat straw is clearer as the same open and closed symbol is used (square) and other treatments can be compared with the two.
- Describing symbols in the figure legend instead of using an inserted legend leaves more white space to help readers compare the lines.
- The x axis is more descriptively titled and units are more appropriately spaced.
- The title has changed from a descriptive noun phrase to a story-telling statement of what the data show.

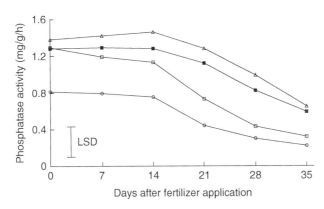

Fig. AP.1 Root surface phosphatase activity of wheat plants differed after soil amendment with different fertilizer treatments. Phosphatase activity was highest in farmyard manure (△) treatments, followed by combined application of chemical fertilizer and wheat straw (■), chemical fertilizer alone (□), and control/no amendment (○) treatments. Phosphatase activity declined over 5 weeks for all treatments. Least significant difference (LSD; two-way ANOVA, $P \leq 0.05$) is 0.39 mg/g/h.

Task 5.3 Improving information visualisation in a table

Improvements in Table AP.1 are as follows:

- The title has changed from a description to a story-telling statement of what the data show.
- Sampled soils are sorted to better highlight the correlated gradient of clay content and K concentration within them (in the original table, the soils are presented in the order in which they were collected, not the order in which they highlight correlation). The soil samples could be renamed to present them in the new order.

Table AP.1 Soil texture correlates with K concentration determined using three extraction methods: WS = water soluble, $CaCl_2$ = calcium chloride, NaTPB = sodium tetraphenyl boron (SD = standard deviation).

			mg K/kg soil		
Soil	Clay (g/kg)	Silt (g/kg)	WS	$CaCl_2$	NaTPB
7	360	190	6	34	360
1	380	200	10	41	480
3	410	230	15	57	583
4	434	205	19	70	652
8	440	235	20	87	723
5	485	235	27	100	932
2	535	265	31	162	1208
6	610	282	50	290	1730
Mean	457	230	22	105	834
SD (±)	83	32	14	85	449

- Mean and standard deviation values have been rounded back (which prevents presentation of false accuracy and reduces clutter).
- A small break between the data columns and the mean values improves the visual appreciation of the gradient of soil texture and K concentrations.

Task 5.5 Identifying parts of figure legends

Parts of the legend from Kaiser et al. (2003)

Sentence	Part
Uptake of Fe(II) by GmDmt1 in yeast.	Part 1
(a) Influx of $^{55}Fe^{2+}$ into yeast cells transformed with GmDmt1;1, *fet3fet4* cells	Part 1
were transformed with GmDmt1;1-pFL61 or pFL61 and then incubated with $1 \mu M$ $^{55}FeCl_3$ (pH 5.5) for 5- and 10-min periods.	Part 3
Data presented are means ± SE of ^{55}Fe uptake between 5 and 10 min from three separate experiments (each performed in triplicate).	Part 4
(b) Concentration dependence of ^{55}Fe influx into *fet3fet4* cells transformed with GmDmt1;1-pFL61 or pFL61.	Part 1
Data presented are means ± SE of ^{55}Fe uptake over 5 min ($n = 3$).	Part 4
The curve was obtained by direct fit to the Michaelis-Menten equation.	Part 2
Estimated K_M and V_{MAX} for GmDmt1;1 were $6.4 \pm 1.1 \mu M$ Fe(III) and 0.72 ± 0.08 nM Fe(III)/min/mg protein, respectively.	Part 2
(c) Effect of other divalent cations on uptake of $^{55}Fe^{2+}$ into *fet3fet4* cells transformed with pFL61-GmDMT1;1.	Part 1
Data presented are means ± SE of ^{55}Fe ($10 \mu M$) uptake over 10 min in the presence and absence of $100 \mu M$ unlabelled Fe^{2+}, Cu^{2+}, Zn^{2+} and Mn^{2+}.	Parts 3 and 4

Source: Kaiser, B.N., Moreau, S., Castelli, J., et al. The soybean NRAMP homologue, GmDMT1, is a symbiotic divalent metal transporter capable of ferrous iron transport. The Plant Journal, 35, 295–304. © 2003, John Wiley & Sons.

Answer pages

Answer pages

Sentence	Part
Number of *Sargassum muticum* (a) recruits and (b) adults in field experiment plots (900 cm²).	Part 1
Propagule pressure is grams of reproductive tissue suspended over experimental plots at beginning of experiment.	Part 3
The average mass of an adult *S. muticum* (174 g) is indicated by an arrow.	Part 5
Data are means ± 1 SE (*n* = 3).	Part 4

Source: Britton-Simmons, K.H. & Abbott, K.C. Short- and long-term effects of disturbance and propagule pressure on a biological invasion. Journal of Ecology, 96, 68–77. © 2008, John Wiley & Sons.

Parts of the legend from Ganci et al. (2012)

Sentence	Part
A sequence of SEVIRI images recorded from 9:15 to 10:30 GMT on 13 May.	Part 1
At 9:15 GMT, no evident eruptive phenomena are observable.	Part 3
At 9:30 GMT the image shows the beginning of an ash plume (yellow pixels) moving toward NE, which was associated to the lava fountain from the northern part of the eruptive fissure.	Part 3
The first hot spot, due to the increased lava output and thermal anomaly, was detected by HOTSAT at 10:30 GMT.	Part 3

Source: Ganci, G., Vicari, A., Cappello, A., & Del Negro, C. An emergent strategy for volcano hazard assessment: from thermal satellite monitoring to lava flow modelling. Remote Sensing of Environment, 119, 197–207. © 2012, Elsevier.

Task 6.1 Separate location sentences in Results sections

Kaiser et al. (2003): no separate location sentences occur.

Britton-Simmons & Abbott (2008): only one separate location sentence occurs, and it is written in the style of a highlight sentence:

> We plotted the proportion of plots in each treatment combination that were successfully invaded as a function of propagule pressure (Fig. 3).

We can suggest two possible reasons for this choice:

1 this style allows the use of the active voice verb, in line with the more direct writing style preferred by these authors; or
2 the sentence is part of a longer section detailing what was done to answer the question posed at the start of the paragraph. The style of the sentence fits well with the way the other sentences have been constructed.

Ganci et al. (2012): In Section 5, a single sentence appears which has only a "locate" function, and the location element (*Fig. 8*) appears at the end of the sentence as a result of the authors' choice to use a verb in the passive voice (*are shown*). This style is an effective way to highlight what is to be seen in the location more than the location itself:

> The actual flow areas and two simulated scenarios, obtained by using the minimum and maximum derived TADR, are shown in Fig. 8.

Task 7.1 Materials and methods organisation

Question	Kaiser et al. (2003)	Britton-Simmons & Abbott (2008)	Ganci et al. (2012)
1. What subheadings are used in the section?	Experimental procedures; Plant growth; Isolation of GmDmt1;1; Northern analysis; Antibody generation and Western immunoblot analysis; Symbiosome isolation and nodule membrane purification	Methods; Study system; The invader; Field experiment; Statistical analysis; Model	HOTSAT satellite monitoring system; Hot-spot detection; Heat flux and TADR conversion
2.i. How do the subheadings relate to the end of the Introduction?	No specific relationship seen.	Very clear relation to the last paragraph of the Introduction. Wordings related to each subheading have been used there in describing the principal activity of the study, in almost the same order as the subheadings.	Specific relationship: HOTSAT was mentioned at the end of the Introduction as "recently developed" by the researchers and as a component of the new system presented in the paper.
2.ii. How do the subheadings relate to the subheadings in the Results section?	Results subheadings are not specifically related to Experimental procedure subheadings, but the order of the information in the Experimental procedure section follows closely the order in which the results are presented within that section.	The last three subheadings come in the same order in which the Results are presented.	No specific relationship seen.
3. Is the section easy for you to follow? Why? Or why not?	Aids to clarity include frequent use of subheadings relating to the order of the information in Results and use of purpose phrases to show why steps were taken in relation to the experimental aims.	Aids to clarity include overview sentences at the start of paragraphs, before details are given.	Aids to clarity include frequent use of introductory phrases with "to + verb" to show why actions were taken.

Examples of transformed sentences:

PEA	Original sentence	Transformation
Kaiser et al. (2003)	Soybean seeds were inoculated at planting with *Bradyrhizobium japonicum* USDA 110 . . . [passive]	We inoculated soybean seeds at planting with *Bradyrhizobium japonicum* USDA 110 . . . [active]
	Subsequent PCR experiments identified a full-length 1849-bp cDNA . . . [active]	A full-length 1849-bp cDNA was identified in subsequent PCR experiments . . . [passive]
Britton-Simmons & Abbott (2008)	Control plots were not altered in any way . . . [passive]	We did not alter control plots in any way . . . [active]
	Each holdfast produces as many as 18 laterals in the early spring . . . [active]	As many as 18 laterals are produced by each holdfast in early spring . . . [passive]
Ganci et al. (2012)	We integrated the HOTSAT system and the MAGFLOW model using . . . [active]	The HOTSAT system and the MAGFLOW model were integrated using . . . [passive]
	The satellite-derived discharge rate estimates are then used in the scenario forecasting module . . . [passive]	We then use the satellite-derived discharge rate estimates in the scenario forecasting module . . . [active][a]

[a] This transformation changes the meaning to one that is not accurate: it is not the researchers who use the estimates, but rather the software. This sentence exemplifies a situation where the passive voice is not only preferred but almost essential.

Task 7.4 Top-heavy passive sentences

Improved versions:

The soil water balance equation (Xin, 1986; Zhu and Niu, 1987) was used to compute actual evapotranspiration (T) for each crop, defined as the amount of precipitation for the period between sowing and harvesting the particular crop plus or minus the change in soil water storage in the 2 m soil profile.

or:

Actual evapotranspiration (T) for each crop was computed by the soil water balance equation (Xin, 1986; Zhu and Niu, 1987). This measure is defined as the amount of precipitation for the period between sowing and harvesting the particular crop plus or minus the change in soil water storage in the 2 m soil profile.

Task 8.1 Introduction stages

Stages in the Introduction of Kaiser et al. (2003)

Text	Stage
Legumes form symbiotic associations with N_2-fixing soil-borne bacteria of the *Rhizobium* family. The symbiosis begins when compatible bacteria invade legume root hairs, signalling the division of inner cortical root cells and the formation of a nodule. Invading bacteria migrate to the developing nodule by way of an 'infection thread', comprised of an invaginated cell wall. In the inner cortex, bacteria are released into the cell cytosol, enveloped in a modified plasma membrane (the peribacteroid membrane (PBM)), to form an organelle-like structure called the symbiosome, which consists of bacteroid, PBM and the intervening peribacteroid space (PBS; Whitehead and Day, 1997). The bacteria, subsequently, differentiate into the N_2-fixing bacteroid form. The symbiosis allows the access of legumes to atmospheric N_2, which is reduced to NH_4^+ by the bacteroid enzyme nitrogenase. In exchange for reduced N, the plant provides carbon to the nodules to support bacterial respiration, a low-oxygen environment in the nodule suitable for bacteroid nitrogenase activity, and all the essential nutritional elements necessary for bacteroid activity. Consequently, nutrient transport across the PBM is an important control mechanism in the promotion and regulation of the symbiosis.	Stage 1 (providing a context for the problem to be investigated)
Micronutrients such as iron are essential for bacteroid activity and nodule development. The demand for iron increases during symbiosis (Tang et al., 1990), where the metal is used for the synthesis of various iron-containing proteins in both the plant and the bacteroids. In the plant fraction, iron is an important part of the heme moiety of leghemoglobin, which facilitates the diffusion of O_2 to the symbiosomes in the affected cell cytosol (Appleby, 1984). In bacteroids, there are many iron-containing proteins involved in N_2 fixation, including nitrogenase itself and cytochromes used in the bacteroid electron-transport chain. In the soil, iron is often poorly available to plants as it is usually in its oxidised form Fe(III), which is highly insoluble at neutral and basic pH. To compensate this, plants have developed two general strategies to gain access to iron from their localised environment. Strategy I involves secretion of phytosiderophores that aid in the solubilisation and uptake of Fe(III), while Strategy II involves initial reduction of Fe(III) to Fe(II) by a plasma membrane Fe(III)-chelate reductase, followed by uptake of Fe(II) (Romheld, 1987).	Stage 3 (broad research niche, claiming importance) Stage 1 (another aspect of the context, iron, as indicated in the title)
The mechanism(s) involved in bacteroid iron acquisition within the nodule have been investigated at the biochemical level, and three activities have been identified (Day et al., 2001). Fe(III) is transported across the PBM complexed with organic acids such as citrate, and accumulates in the PBS (Levier et al., 1996; Moreau et al., 1995), where it becomes bound to siderophore-like compounds (Wittenberg et al., 1996). Fe(III) chelate reductase activity has been measured on isolated PBM, and Fe(III) uptake into isolated symbiosomes is stimulated by Nicotinamide Adenine Dinucleotide (NADH), reduced form (Levier et al., 1996). However, Fe(II) is also readily transported across the PBM and has been found to be the favoured form of iron taken up by bacteroids (Moreau et al., 1998). The proteins involved in this transport have not yet been identified.	Stage 2 (aspects of the problem already investigated by others) Stage 3 (more specific research gap)

Two classes of putative Fe(II)–transport proteins (Irt/Zip and Dmt/Nramp) have been identified in plants (Belouchi et al., 1997; Curie et al., 2000; Eide et al., 1996; Thomine et al., 2000). The Irt/Zip family was first identified in *Arabidopsis* by functional complementation of the yeast Fe(II) transport mutant DEY1453 (*fet3fet4*; Eide et al., 1996). *Atlrt1* expression is enhanced in roots when grown on low iron (Edie et al., 1996) and appears to be the main avenue for iron acquisition in *Arabidopsis* (Vert et al., 2002). Recently a soybean Irt/Zip isologue, GmZip1, was identified and localised to the PBM in nodules (Moreau et al., 2002). GmZip1 has been characterised as a symbiotic zinc transporter, which does not transport Fe(II). The second class of iron-transport proteins consists of the Dmt/Nramp family of membrane transporters, which were first identified in mammals as a putative defense mechanism utilised by macrophages against mycobacterium infection (Supek et al., 1996; Vidal and Gros, 1994). Mutations in Nramp proteins in different organisms result in varied phenotypes including altered taste patterns in *Drosophila* (Rodrigues et al., 1995), microcyticanaemia (mk) in mice and Belgrade rats (Fleming et al., 1997) and loss of ethylene sensitivity in plants (Alonso et al., 1999). The rat and yeast NRAMP homologues (DCT1 and SMF1, respectively) have been expressed in Xenopus oocytes and shown to be broad-specificity metal ion transporters capable of Fe(II), among other divalent cations, transport (Chen et al., 1999; Gunshin et al., 1997). The plant homologue, AtNramp1, complements the growth defect of the yeast Fe(II) transport mutant DEY1453, while other *Arabidopsis* members do not (Curie et al., 2000; Thomine et al., 2000). Interestingly, AtNramp1 overexpression in *Arabidopsis* also confers tolerance to toxic concentrations of external Fe(II) (Curie et al., 2000), suggesting, perhaps, that it is localised intracellularly.

In this study we have identified a soybean homologue of the Nramp family of membrane proteins, GmDmt1;1. We show that GmDmt1;1 is a symbiotically enhanced plant protein, expressed in soybean nodules at the onset of nitrogen fixation, and is localised to the PBM. GmDmt1;1 is capable of Fe(II) transport when expressed in yeast. Together, the localisation and demonstrated activity of GmDmt1;1 in soybean nodules suggests that the protein is involved in Fe(II) transport and iron homeostasis in the nodule to support symbiotic N_2 fixation.

Stage 2

Stage 3 (building the gap)

Stage 2

Stage 4 (principal activity of the study and conclusion reached)

Source: Kaiser, B.N., Moreau, S., Castelli, J., et al. The soybean NRAMP homologue, GmDMT1, is a symbiotic divalent metal transporter capable of ferrous iron transport. The Plant Journal, 35, 295–304. © 2003, John Wiley & Sons.

Stages in the Introduction of Britton-Simmons & Abbott (2008)

Text	Stage
Biological invasions are a global problem with substantial economic (Pimentel *et al.* 2005) and ecological (Mack *et al.* 2000) costs. Research on invasions has provided important insights into the establishment, spread and impact of non-native species. One key goal of invasion biology has been to identify the factors that determine whether an invasion will be successful (Williamson 1996). Accordingly, ecologists have identified several individual factors (e.g. disturbance and propagule pressure) that appear to exert strong controlling influences on the invasion process. However, understanding how these processes interact to regulate invasions remains a major challenge in ecology (D'Antonio et al. 2001; Lockwood et al. 2005; Von Holle & Simberloff 2005).	Stage 1
	Stage 3 (broad research niche, claiming significance)
Propagule pressure is widely recognized as an important factor that influences invasion success (MacDonald *et al.* 1989; Simberloff 1989; Williamson 1996; Lonsdale 1999; Cassey *et al.* 2005). Previous studies suggest that the probability of a successful invasion increases with the number of propagules released (Panetta & Randall 1994; Williamson 1989; Grevstad 1999), with the number of introduction attempts (Veltman *et al.* 1996), with introduction rate (Drake *et al.* 2005), and with proximity to existing populations of invaders (Bossenbroek *et al.* 2001). Moreover, propagule pressure may influence invasion dynamics after establishment by affecting the capacity of non-native species to adapt to their new environment (Ahlroth *et al.* 2003; Travis *et al.* 2005). Despite its acknowledged importance, propagule pressure has rarely been manipulated experimentally and the interaction of propagule pressure with other processes that regulate invasion success is not well understood (D'Antonio et al. 2001; Lockwood et al. 2005).	Stage 2
	Stage 3 (one component of the study, as indicated in the title)
Resource availability is a second key factor known to influence invasion success and processes that increase or decrease resource availability therefore have strong effects on invasions (Davis *et al.* 2000). Resource pre-emption by native species generates biotic resistance to invasion (Stachowicz *et al.* 1999; Naeem *et al.* 2000; Levine *et al.* 2004). Consequently, physical disturbance can facilitate invasions by reducing competition for limiting resources (Richardson & Bond 1991; Hobbs & Huenneke 1992; Kotanen 1997; Prieur-Richard & Lavorel 2000). In most communities disturbances occur via multiple mechanisms and the disturbances created by different agents vary in their intensity and frequency (D'Antonio *et al.* 1999). Recent empirical (Larson 2003; Hill *et al.* 2005) and theoretical (Higgins & Richardson 1998) studies suggest that not all types of disturbance have equivalent effects on the invasion process.	Stage 2
Moreover, most of what we know about the effects of disturbance on invasions comes from short-term experimental studies. It is presently unclear how different disturbance agents influence long-term patterns of invasion.	Stage 3 (another component, as highlighted in the title)

In order for any invasion to be successful, propagule arrival must coincide with the availability of resources needed by the invading species (Davis *et al.* 2000). Therefore, the interaction between propagule pressure and processes that influence resource availability will ultimately determine invasion success (Brown & Peet 2003; Lockwood *et al.* 2005; Buckley *et al.* 2007).

Stage 2

In this study we used the invasion of shallow, subtidal kelp communities in Washington State by the Japanese seaweed Sargassum muticum as a study system to better understand the effects of propagule pressure and disturbance on invasion. In a factorial field experiment we manipulated both propagule pressure and disturbance in order to examine how these factors independently and interactively influence S. muticum establishment in the short term. We supplement the experimental results with a parameterized integrodifference equation model, which we use to examine how different natural disturbance agents influence the spread of S. muticum through the habitat in the longer term. Although a successful invasion clearly requires both establishment and spread of the invader, most studies have looked at just one of these processes (Melbourne *et al.* 2007). We take an integrative approach by employing both a short-term experiment and a longer-term model, allowing us to examine the effects of disturbance and propagule limitation on the entire invasion process.

Stage 4 (principal activities of the present study)

Stage 5 (value of the present study, claiming significance)

Source: Britton-Simmons, K.H. & Abbott, K.C. Short- and long-term effects of disturbance and propagule pressure on a biological invasion. Journal of Ecology, 96, 68–77. © 2008, John Wiley & Sons.

Stages in the Introduction of Ganci et al. (2012)

Text	Stages
Etna volcano, Italy, is characterized by persistent activity, consisting of degassing and explosive phenomena at its summit craters, as well as frequent flank eruptions. All eruption typologies can give rise to lava flows, which are the greatest hazard presented by the volcano to inhabited and cultivated areas (Behncke et al., 2005). The frequent eruptions of Mt Etna and hazard they pose represent an excellent opportunity to apply both spaceborne remote sensing techniques for thermal volcano monitoring, and numerical simulations for predicting the area most likely to be inundated by lava during a volcanic eruption. Indeed, Etna has witnessed many tests involving the application of space-based remote sensing data to detect, measure and track the thermal expression of volcanic effusive phenomena (e.g. Ganci et al., 2011b; Harris et al., 1998; Wright et al., 2004), as well as a number of lava flow emplacement models (e.g. Crisci et al., 1986; Harris & Rowland, 2001; Vicari et al., 2007). These efforts have allowed key at-risk areas to be rapidly and appropriately identified (Ganci et al., 2011b; Wright et al., 2008).	Stage 1
Over the last 25 years, satellite measurements in the thermal infrared have proved well suited to detection of volcanic thermal phenomena (e.g. Francis & Rothery, 1987; Harris et al., 1995) and to map the total thermal flux from active lava flows (e.g. Flynn et al., 1994; Harris et al., 1998). High spatial resolution data collected by Landsat and ASTER have been employed for the thermal analysis of active lava flows (Hirn et al., 2007; Oppenheimer, 1991), lava domes (Carter & Ramsey, 2010; Kaneko et al., 2002; Oppenheimer et al., 1993), lava lakes (Harris et al., 1999; Wright et al., 1999), and fumarole fields (Harris & Stevenson, 1997; Pieri & Abrams, 2005). Lower spatial, but higher temporal, resolution sensors, such the Advanced Very High Resolution Radiometer (AVHRR) and the MODerate Resolution Imaging Spectro-radiometer (MODIS), have also been widely used for infrared remote sensing of volcanic thermal features, as has GOES (e.g. Rose & Mayberry, 2000). The high temporal resolution (15 min) offered by the Spinning Enhanced Visible and Infrared Imager (SEVIRI), already employed for the thermal monitoring of effusive volcanoes in Europe and Africa (Hirn et al., 2008), has recently been exploited to estimate lava discharge rates for eruptive events of short duration (a few hours) at Mt Etna (Bonaccorso et al., 2011a, 2011b). These data have also proved capable of detecting, measuring and tracking volcanic thermal phenomena, despite the fact that the volcanic thermal phenomena are usually much smaller than the nominal pixel sizes of 1–4 km (Ganci et al., 2011a, 2011b; Vicari et al., 2011b). Great advances have also been made in understanding the physical processes that control lava flow emplacement, resulting in the development of a range of tools allowing the assessment of lava flow hazard. Many methods have been developed to simulate and predict lava flow paths and run-out distance based on various simplifications of the governing physical equations and on analytical and empirical modeling (e.g.: Hulme, 1974; Crisci et al., 1986; Young & Wadge, 1990; Miyamoto & Sasaki, 1997; Harris & Rowland, 2001; Costa & Macedonio, 2005; Del Negro et al., 2005). In its most simple form, such forecasting may involve the application of volcano-specific empirical length/effusion rate relationships to estimate flow length (e.g. Calvari & Pinkerton, 1998). At their most complex, such predictions may involve iteratively solving a system of equations that characterize the effects that cooling-induced changes in rheology have on the ability of lava to flow downhill, while taking into account spatial variations in slope determined from a digital elevation model (e.g. Del Negro et al., 2008).	Stage 2

On Etna, the MAGFLOW Cellular Automata model has successfully been used to reproduce lava flow paths during the 2001, 2004 and 2006 effusive eruptions (Del Negro et al., 2009; Herault et al., 2008; Vicari et al., 2007). More recently, it has been applied to consider the impact of hypothetical protective barrier placement on lava flow diversion (Scifoni et al., 2010) and has been used for the production of a hazard map for lava flow invasion on Etna (Cappello et al., 2011a, 2011b). MAGFLOW is based on a physical model for the thermal and rheological evolution of the flowing lava. To determine how far lava will flow, MAGFLOW requires constraint of many parameters. However, after chemical composition, the instantaneous lava output at the vent is the principal parameter controlling the final dimensions of a lava flow (e.g. Harris & Rowland, 2009; Pinkerton & Wilson, 1994; Walker, 1973). As such, any simulation technique that aims to provide reliable lava flow hazard assessments should incorporate temporal changes in discharge rate into its predictions in a timely manner. Satellite remote sensing provides a means to estimate this important parameter in real-time during an eruption (e.g. Harris et al., 1997; Wright et al., 2001). A few attempts to use satellite-derived discharge rates to drive numerical simulations have already been made (e.g. Herault et al., 2009; Vicari et al., 2009; Wright et al., 2008). These modeling efforts all used infrared satellite data acquired by AVHRR and/or MODIS to estimate the time-averaged discharge rate (TADR) following the methodology of Harris et al. (1997). Recently, we developed the HOTSAT multiplatform system for satellite infrared data analysis, which is capable of managing multispectral data from different sensors as MODIS and SEVIRI to detect volcanic hot spots and output their associated radiative power (Ganci et al., 2011b). Satellite-derived output for lava flow modeling purposes has already been tested for the 12–13 January 2011 paroxysmal episode at Mt Etna by Vicari et al., 2011b. Here we present a new web-based GIS framework, named LAV@HAZARD, which integrates the HOTSAT system for satellite-derived discharge rate estimates with the MAGFLOW model to simulate lava flow paths. As a result, LAV@HAZARD now represents the central part of an operational monitoring system that allows us to map the probable evolution of lava flow-fields while the eruption is ongoing. Here we describe and demonstrate the operation of this LAV@HAZARD using a retrospective analysis of Etna's 2008–2009 flank eruption. This eruption was exceptionally well documented by a variety of monitoring techniques maintained by INGV-CT including spaceborne thermal infrared measurements. Within this operational role, HOTSAT is first used to identify thermal anomalies due to active lava flows and to compute the TADR. This is then used to drive MAGFLOW simulations, allowing us to effectively simulate the advance rate and maximum length for the active lava flows. Because satellite-derived TADRs can be obtained in real times and simulations spanning several days of eruption can be calculated in a few minutes, such a combined approach has the potential to provide timely predictions of the areas likely to be inundated with lava, which can be updated in response to changes of the eruption conditions as detected by the current image conditions. If SEVIRI data are used, simulations can be updated every 15 min. Our results thus demonstrate how LAV@HAZARD can be exploited to produce realistic lava flow hazard scenarios and for helping local authorities in making decisions during a volcanic eruption.

Stage 3 (need/problem to be addressed)

Stage 1/2; Stage 3 implicit in "a few"

Stage 4 (main activity of the paper)

Stage 2
Stage 4

Stage 5

Source: Ganci, G., Vicari, A., Cappello, A., & Del Negro, C. An emergent strategy for volcano hazard assessment: from thermal satellite monitoring to lava flow modelling. Remote Sensing of Environment, 119, 197–207. © 2012, Elsevier.

Answer pages

Task 8.2 Introduction Stage 1 analysis

Question	Kaiser et al. (2003)	Britton-Simmons & Abbott (2008)	Ganci et al. (2012)
Are some sentences written in the present tense? How many?	Yes, 8	Yes, 2	Yes, 3
Are some sentences written in the present perfect tense? How many?	No	Yes, 3	Yes, 2
Which tense is used more? Why do you think this is the case?	Present, because the focus of the content is explaining a biological process	Present perfect, because the focus is on the developing field of research and the work others have done up to the present	Present, because the focus is on describing the features of the study site that make the study important
How many sentences contain references?	1 (of 8)	3 (of 5)	3 (of 5)
What kinds of sentences do not have references?	Sentences that summarise commonly accepted knowledge in the field	Sentences that summarise the current state of knowledge in the field	Sentences that state facts and views commonly accepted in the field

Task 8.3 "Country" to "city" in Stage 1

Kaiser et al. (2003)

Country: Legume symbiotic associations.
Province: The peribacteroid membrane (PBM) and its role.
City: Nutrient transport across the PBM.

Britton-Simmons & Abbott (2008)

Country: Biological invasions.
Province: Factors controlling the invasion process.
City: The interaction of the factors and processes.

Ganci et al. (2012)

Country: Etna volcano.
Province: Hazard caused by lava flow from volcanoes.
City: Application of relevant technologies for hazard identification.

Old information is underlined:

Legumes form symbiotic associations with N_2-fixing soil-borne bacteria of the *Rhizobium* family. <u>The symbiosis</u> begins when compatible bacteria invade legume root hairs, signalling the division of inner cortical root cells and the formation of a nodule. <u>Invading bacteria</u> migrate to the developing nodule by way of an "infection thread," comprised of an invaginated cell wall. In the inner cortex, <u>bacteria</u> are released into the cell cytosol, enveloped in a modified plasma membrane (the peribacteroid membrane (PBM)), to form an organelle-like structure called the symbiosome, which consists of bacteroid, PBM and the intervening peribacteroid space (PBS; Whitehead and Day, 1997). <u>The bacteria</u>, subsequently, differentiate into the N_2-fixing bacteroid form. <u>The symbiosis</u> allows the access of legumes to atmospheric N_2, which is reduced to NH_4^+ by the bacteroid enzyme nitrogenase. In exchange for <u>reduced N</u>, the plant provides carbon to the nodules to support bacterial respiration, a low-oxygen environment in the nodule suitable for bacteroid nitrogenase activity, and all the essential nutritional elements necessary for bacteroid activity. Consequently, <u>nutrient transport</u> across the PBM is an important control mechanism in the promotion and regulation of the symbiosis.

Task 8.5 Identifying plagiarism

Plagiarised sentence in Version 2	Reason for the problem
However, this technique is not adaptable to all plants, particularly pasture species.	This sounds like the idea of the writer of the paragraph, but we know from Version 1 that it was originally the idea of Russell & Fillery (1996). Because there is no grammatical link between the two sentences, the reference in the first sentence does not apply to the second. Note in Version 1 that the authors used both a grammatical link (*they*) and a tense marker (past tense *was not adaptable*) to indicate that the idea came from the cited work.

Task 8.6 Signal words for the research gap or niche

McNeill et al. (1997)	Kaiser et al. (2003)	Britton-Simmons & Abbott (2008)	Ganci et al. (2012)
scarce, with little account taken of, is accordingly required, but, however	consequently, however, have not yet been identified, putative, appears to be	remains a major challenge, despite its acknowledged importance, rarely, is not well understood, it is presently unclear how, to better understand	should incorporate, in a timely manner

Task 8.8 Stage 4 sentence templates

McNeill et al. (1997)

The experiments reported here were designed (i) to assess the use of [NP1] to [verb phrase], and (ii) to obtain quantitative estimates of [NP2].

Kaiser et al. (2003)

In this study we have identified [NP1], [NP2]. We show that [NP2] is [NP3], expressed in [NP4] at the onset of [NP5], and is localised to [NP6].

Britton-Simmons & Abbott (2008)

In this study we used [NP1] as a study system to better understand the effects of [NP2] and [NP3] on [NP4]. In a [adjectives] experiment we manipulated both [NP2] and [NP3] in order to examine how these factors [adverbs] influence [NP5] in the short term. We supplement the experimental results with [NP6], which we use to examine how different [NP7] influence [NP8] in the longer term.

Ganci et al. (2012)

Here we present [NP1], which integrates [NP2] with [NP3] to simulate [NP4] . . . Here we describe and demonstrate the operation of this [NP1] using a [adjective] analysis of [NP5].

Task 8.10 Topic sentence analysis

Reference	Topic sentence	Previous paragraph	Upcoming paragraph
Kaiser et al. (2003)	Two classes of putative Fe(II)-transport proteins (Irt/Zip and Dmt/Nramp) have been identified in plants (references).	Ends by stating that the proteins involved are *unknown*, which links directly to *putative* (= possible candidates) in this sentence.	Gives details of research results on each of the two classes, in the same order in which they are referred to in the topic sentence (Irt/Zip and then Dmt/Nramp).
Britton-Simmons & Abbott (2008)	Propagule pressure is widely recognized as an important factor that influences invasion success (references).	Refers to *propagule pressure* as one of two examples of factors influencing invasions.	Gives details of results of previous studies showing ways in which *propagule pressure* affects invasion success.
Ganci et al. (2012)	On Etna, the MAGFLOW Cellular Automata model has successfully been used to reproduce lava flow paths during the 2001, 2004 and 2006 effusive eruptions (references).	Describes many methods and models used to simulate and predict lava flows.	Reports several successful uses of the MAGFLOW model and introduces the constraints to the use of this and other models that are to be addressed in the present study.

Problem sentence, with inappropriately located new information underlined:

A sudden increase in the number of sick and coughing pigs and a sharp rise in mortalities among grower/finisher pigs may herald an outbreak of APP in a herd.

Revised wording:

An outbreak of APP in a herd may be heralded by a sudden increase in the number of sick and coughing pigs and a sharp rise in mortalities among grower/finisher pigs.

N.B. Making this improvement involves changing an active voice verb (*may herald*) to a passive voice verb (*may be heralded*). This is one of the useful features of the active/passive verb system in English: it allows us to change the order of the information in a sentence. For science writers, this means we have an extra tool to enable us to get the given information at the start of each sentence.

Task 8.12 Revising top-heavy sentences

1 Original: In this project the *Rhizoctonia* populations of two field soils in the Adelaide Plains region of South Australia were characterised.

Suggested improvement options:

This project characterised the *Rhizoctonia* populations of two field soils in the Adelaide Plains region of South Australia.

or:

The aim of this project was to characterise the *Rhizoctonia* populations of two field soils in the Adelaide Plains region of South Australia.

2 Original: A balance between deep and shallow rooting plants, heavy and light feeders, nitrogen fixers and consumers and an undisturbed phase is needed to achieve maximum benefit through rotation.

Suggested improvement options:

Maximum benefit through rotation can be achieved by using a balance between deep and shallow rooting plants, heavy and light feeders, nitrogen fixers and consumers and an undisturbed phase.

or:

To achieve maximum benefit through rotation, it is necessary to have a balance between deep and shallow rooting plants, heavy and light feeders, nitrogen fixers and consumers and an undisturbed phase.

Task 9.2 Information elements in the Discussion section

Information elements in the Discussion of Kaiser et al. (2003)

Sentences	Information element
The competition experiments shown in Figure 5(c) indicate that GmDmt1 can transport other divalent cations in addition to ferrous iron. Zinc, copper and manganese all inhibited iron uptake. The ability of GmDmt1;1 to enhance growth of the *zrt1zrt2* yeast mutant further suggests that the protein is not specific for iron transport. The preferred substrate *in vivo;* may well depend on the relevant concentrations of divalent metals in the infected cell cytosol. This lack of specificity has been found with Nramp homologues from other organisms, including Nramp2 from mice. Despite this lack of specificity when expressed in heterologous systems, mutation of murine Nramp2 results in an anaemic phenotype, demonstrating that *in vivo;* it is predominantly an iron transporter (Fleming *et al.*, 1997).	2a. Restatement of one of the main findings, showing how it contributes to the main activity of the study 3. Speculation about the finding 2b. Comparison with the findings of other researchers
Although GmDmt1;1 was able to complement the DEY1453 (*fet3fet4*) yeast mutant, the complementation was not robust and the growth media had to be supplemented with low concentrations of iron. Atlrt1, on the other hand, showed much better complementation and allowed growth of the mutant in the absence of added iron (Figure 4). There are several possible reasons for the poorer growth with GmDmt1;1, including possible instability of GmDmt1;1 transcripts (perhaps because of the presence of the regulatory IRE element in the transcript).	2a. Continued review of the finding 3. Speculation about the finding

Source: Kaiser, B.N., Moreau, S., Castelli, J., et al. The soybean NRAMP homologue, GmDMT1, is a symbiotic divalent metal transporter capable of ferrous iron transport. The Plant Journal, 35, 295–304. © 2003, John Wiley & Sons.

Information elements in the Discussion of Britton-Simmons & Abbott (2008)

Sentences	Information elements
Our experimental results demonstrate that space- and propagule-limitation both regulate *S. muticum* recruitment. Our finding that *S. muticum* recruitment was positively related to propagule input is similar to those of two previous studies (Parker 2001; Thomsen *et al.* 2006), in which the propagule input of invasive plants was manipulated. In our control treatment space was limiting, a result that has also been found in previous studies of *S. muticum* recruitment (Deysher & Norton 1982; De Wreede 1983; Sanchez & Fernandez 2006).	2a. Restatement of the most important finding, showing how it contributes to the main activity of the study 2b. Comparisons with the findings of other researchers
Consequently, increasing propagule pressure had a relatively weak effect on recruitment in undisturbed plots (Fig. 1a). However, when space limitation was alleviated by disturbing the plots, increasing propagule pressure caused a dramatic	2a. Continued review of the important findings

increase in recruitment (Fig. 1a). This suggests that in the presence of adequate substratum for settlement, propagule limitation becomes the primary factor controlling *S. muticum* recruitment. These results indicate that *S. muticum* recruitment under natural field conditions will be determined by the interaction between disturbance and propagule input.

5. Implications of the results (what they mean in the context of the broader field)

Source: Britton-Simmons, K.H. & Abbott, K.C. Short- and long-term effects of disturbance and propagule pressure on a biological invasion. Journal of Ecology, 96, 68–77. © 2008, John Wiley & Sons.

Information elements in the Discussion of Ganci et al. (2012)

Sentences	Information elements
The uncertainty in satellite derived TADR estimates is quite large, up to about 50%, but it is comparable to the error in field-based effusion rate measurements (Calvari et al., 2003; Harris & Baloga, 2009; Harris & Neri, 2002; Harris et al., 2007; Sutton et al., 2003). The main uncertainties arise from the lack of constraint on the lava parameters used to convert thermal flux to TADR (Harris & Baloga, 2009). Moreover, the presence of ash strongly affected the satellite-based estimates of lava discharge rate. This led to an underestimation of the satellite-derived final volume, and to a difference in the timing of simulated lava flow emplacement. Moreover, a certain discrepancy between actual and modeled flow areas is to be expected, as changes in the contemporaneous effusion rate take time to be translated into a change in active flow area (that is, the active flow area at any one time is a function of the antecedent effusion rate, rather than the instantaneous effusion rate) (Harris & Baloga, 2009; Wright et al., 2001).	2a and 2b. Restatement of a finding and comparison with the findings of other researchers 3. Explanation of findings, supported by citation 3. Explanations of findings (based on the study conditions) 3. Explanation of findings, supported by citation

Source: Ganci, G., Vicari, A., Cappello, A., & Del Negro, C. An emergent strategy for volcano hazard assessment: from thermal satellite monitoring to lava flow modelling. Remote Sensing of Environment, 119, 197–207. © 2012, Elsevier.

Task 9.5 Negotiating strength of claims with verbs

| The presence of an IRE motif | implies
suggests

provides evidence
indicates
shows
demonstrates | that GmDmt1;1 mRNA | might be stabilized
could be stabilized
may be stabilized
can be stabilized
was stabilized
should be stabilized
is stabilized | by the binding of IRPs in soybean nodules when free iron levels are low. | Weak ↓ Strong |

- The relative strength of the meanings of these terms is to some extent a matter of opinion, and native speakers of English often disagree about the fine detail. The suggestions given represent the outcomes obtained by using this exercise in workshops over a number of years. We have included the exercise to raise your awareness of the issue of negotiating strength. We hope you will pay particular attention to how these words and forms are used in the papers you read in future, with the aim of fine-tuning your understanding of their usage in your own discipline area.
- *Can* indicates that the result has been recorded once, and is therefore possible. It does not make any claim for the likelihood that the result will be repeated.
- *Was stabilized* indicates that no claim is being made that the result is generalisable beyond the conditions of the experiment or study being reported. (In the sentence under consideration, the verb in the next part of the sentence would have to be changed to the past tense also if this alternative were used: *when free iron levels were low*).

3 This usage of *should* indicates strong likelihood, and is often accompanied by a phrase giving the conditions under which the predicted event is likely to occur (as here with *when free iron levels are low*). It is not to be confused with the other usage of *should* to indicate a recommended action (e.g. "You should wash your hands before meals"). The recommended action usage is much less common in scientific writing, although an example does occur in the Conclusions section of Britton-Simmons & Abbott (2008): "The model results demonstrate that caution *should be exercised* when extrapolating the results of short-term disturbance experiments over longer time intervals." This sentence demonstrates an important point: if an author wants to make a recommendation to the reader about future action, the reason for the recommendation needs to be very clear and well-supported. Here, it is the model results, already discussed in detail, that provide the support for making the *should* recommendation.

Task 10.1 Analysing article titles

Question	Kaiser et al. (2003)	Britton-Simmons & Abbott (2008)	Ganci et al. (2012)
Is the title a noun phrase, a sentence, or a question?	Sentence	Noun phrase	Noun phrase
How many words are used in the title?	16	13	15
What is the first idea in the title?	"The soybean NRAMP homologue, GmDMT1": the descriptor and name of the transporter discovered	"Short- and long-term effects"	"An emergent strategy"
Why do you think this idea has been placed first?	The descriptor comes first to show how this new discovery relates to what was previously known about the system under study	This phrase highlights what is new and important about the work being reported	This phrase indicates the status of the work being reported (not finalised)

Analysis of Summary from Kaiser et al. (2003)

Summary sentences	Information elements
Iron is an important nutrient in N_2-fixing legume root nodules. Iron supplied to the nodule is used by the plant for the synthesis of leghemoglobin, while in the bacteroid fraction, it is used as an essential cofactor for the bacterial N_2-fixing enzyme, nitrogenase, and ironcontaining proteins of the electron transport chain. The supply of iron to the bacteroids requires initial transport across the plant-derived peribacteroid membrane, which physically separated bacteroids from the infected plant cell cytosol.	Background
In this study we have identified *Glycine max divalent metal transporter 1 (GmDmt1)*, a soybean homologue of the NRAMP/Dmt1 family of divalent metal ion transporters.	Principal activity
GmDmt1 shows enhanced expression in soybean root nodules and is most highly expressed at the onset of nitrogen fixation in developing nodules.	Results[a]
Antibodies raised against a partial fragment of GmDmt1 confirmed its presence on the peribacteroid membrane (PBM) of soybean root nodules.	Method
GmDmt1 was able to both rescue growth and enhance $^{55}Fe(II)$ uptake in the ferrous iron transport deficient yeast strain *(fet3fet4)*.	Results[a]
The results indicate that GmDmt1 is a nodule-enhanced transporter capable of ferrous iron transport across the PBM of soybean root nodules.	Conclusion
Its role in nodule iron homeostasis to support bacterial nitrogen fixation is discussed.	Another activity of the study/paper

[a] N.B. The first results sentence is written in the present tense: results obtained using the methods employed here are considered to represent outcomes seen as always being true. The second results sentence is written in the past tense; this indicates that the result represents an outcome specific to the experimental conditions used.

Source: Kaiser, B.N., Moreau, S., Castelli, J., et al. The soybean NRAMP homologue, GmDMT1, is a symbiotic divalent metal transporter capable of ferrous iron transport. The Plant Journal, 35, 295–304. © 2003, John Wiley & Sons.

Analysis of Summary from Britton-Simmons & Abbott (2008)

Summary sentences	Information elements
1. Invading species typically need to overcome multiple limiting factors simultaneously in order to become established, and understanding how such factors interact to regulate the invasion process remains a major challenge in ecology.	Background
2. We used the invasion of marine algal communities by the seaweed *Sargassum muticum* as a study system to experimentally investigate the independent and interactive effects of disturbance and propagule pressure in the short term.	Method + principal activity 1
Based on our experimental results, we parameterized an integrodifference equation model, which we used to examine how disturbances created by different benthic herbivores influence the longer term invasion success of *S. muticum*.	Method + principal activity 2

(Continued)

Answer pages

Summary sentences	Information elements
3. Our experimental results demonstrate that in this system neither disturbance nor propagule input alone was sufficient to maximize invasion success. Rather, the interaction between these processes was critical for understanding how the *S. muticum* invasion is regulated in the short term.	Results
4. The model showed that both the size and spatial arrangement of herbivore disturbances had a major impact on how disturbance facilitated the invasion, by jointly determining how much spacelimitation was alleviated and how readily disturbed areas could be reached by dispersing propagules.	Results
5. Synthesis. Both the short-term experiment and the long-term model show that *S. muticum* invasion success is co-regulated by disturbance and propagule pressure. Our results underscore the importance of considering interactive effects when making predictions about invasion success.	Results summary Conclusion/ recommendation

Source: Britton-Simmons, K.H. & Abbott, K.C. Short- and long-term effects of disturbance and propagule pressure on a biological invasion. Journal of Ecology, 96, 68–77. © 2008, John Wiley & Sons.

Analysis of Abstract from Ganci et al. (2012)

Abstract sentences	Information elements
Spaceborne remote sensing techniques and numerical simulations have been combined in a web-GIS framework (LAV@HAZARD) to evaluate lava flow hazard in real time.	Method + Purpose/ principal activity
By using the HOTSAT satellite thermal monitoring system to estimate time-varying TADR (time averaged discharge rate) and the MAGFLOW physics-based model to simulate lava flow paths,	Method
the LAV@HAZARD platform allows timely definition of parameters and maps essential for hazard assessment, including the propagation time of lava flows and the maximum run-out distance.	Results/Conclusion
We used LAV@HAZARD during the 2008–2009 lava flow-forming eruption at Mt Etna (Sicily, Italy). We measured the temporal variation in thermal emission (up to four times per hour) during the entire duration of the eruption using SEVIRI and MODIS data.	Method
The time-series of radiative power allowed us to identify six diverse thermal phases each related to different dynamic volcanic processes and associated with different TADRs and lava flow emplacement conditions.	Result
Satellite-derived estimates of lava discharge rates were computed and integrated for the whole period of the eruption (almost 14 months),	Method
showing that a lava volume of between 32 and 61 million cubic meters was erupted of which about 2/3 was emplaced during the first 4 months.	Result
These time-varying discharge rates were then used to drive MAGFLOW simulations to chart the spread of lava as a function of time.	Method
TADRs were sufficiently low (b30 m³/s) that no lava flows were capable of flowing any great distance so that they did not pose a hazard to vulnerable (agricultural and urban) areas on the flanks of Etna.	Conclusion

Source: Ganci, G., Vicari, A., Cappello, A., & Del Negro, C. An emergent strategy for volcano hazard assessment: from thermal satellite monitoring to lava flow modelling. Remote Sensing of Environment, 119, 197–207. © 2012, Elsevier.

Analysis of Abstract from Harris et al. (2011)

Abstract text	Match with framework stages
Plant development is adapted to changing environmental conditions for optimizing growth. This developmental adaptation is influenced by signals from the environment, which act as stimuli and may include submergence and fluctuations in water status, light conditions, nutrient status, temperature and the concentrations of toxic compounds. The homeodomain-leucine zipper (HD-Zip) I and HD-Zip II transcription factor networks regulate these plant growth adaptation responses through integration of developmental and environmental cues.	Background (information generally available and accepted)
Evidence is emerging that these transcription factors are integrated with phytohormone-regulated developmental networks, enabling environmental stimuli to influence the genetically preprogrammed developmental progression. Dependent on the prevailing conditions, adaptation of mature and nascent organs is controlled by HD-Zip I and HD-Zip II transcription factors through suppression or promotion of cell multiplication, differentiation and expansion to regulate targeted growth. in vitro; assays have shown that, within family I or family II, homo- and/or heterodimerization between leucine zipper domains is a prerequisite for DNA binding. Further, both families bind similar 9-bp pseudopalindromic cis elements, CAATNATTG, under in vitro; conditions.	Results (information representing new synthesis)
However, the mechanisms that regulate the transcriptional activity of HD-Zip I and HD-Zip II transcription factors in vivo are largely unknown. The in planta implications of these protein–protein associations and the similarities in cis element binding are not clear.	Conclusion/recommendation (highlighting remaining gaps in knowledge)

Source: Harris, J.C., Hrmova, M., Lopato, S., & Langridge, P. Modulation of plant growth by HD-Zip class I and II transcription factors in response to environmental stimuli. New Phytologist, 190, 823–837. © 2011, John Wiley & Sons.

Analysis of Abstract from Wymore et al. (2011)

Abstract text	Match with framework stages
Genes and their expression levels in individual species can structure whole communities and affect ecosystem processes.	Background (information generally available and accepted)
Although much has been written about community and ecosystem phenotypes with a few model systems, such as poplar and goldenrod, here we explore the potential application of a community genetics approach with systems involving invasive species, climate change and pollution.	Purpose and scope (including the point of differentiation from previous work)
We argue that community genetics can reveal patterns and processes that otherwise might remain undetected. To further facilitate the community genetics or genes-to-ecosystem concept, we propose four community genetics postulates that allow for the conclusion of a causal	Results (information representing new synthesis/ conclusions)

(Continued)

Abstract text	Match with framework stages
relationship between the gene and its effect on the ecosystem. Although most current studies do not satisfy these criteria completely, several come close and, in so doing, begin to provide a genetic-based understanding of communities and ecosystems, as well as a sound basis for conservation and management practices.	Conclusion/ recommendation (claim for significance of the new synthesis)

Source: Wymore, A.S., Keeley, A.T.H., Yturralde, K.M., Schroer, M.L., Propper, C.R., & Whitham, T.G. Genes to ecosystems: exploring the frontiers of ecology with one of the smallest biological units. New Phytologist, 191 (1), 19–36. © 2011, John Wiley & Sons.

Task 12.2 Analysis of ends of Introductions in reviews

Analysis of end of the Introduction from Harris et al. (2011)

1 Option(s) present in Stage 4: b. Statement of main activity of the paper.
2 Words/phrases identifying the option: We present . . .; We . . . describe. . . and indicate . . . and address the issue of . . .
3 Section functions as map of how article is arranged: Yes (words signalling this function: current knowledge that relates to the roles of the HD-Zip I and HD-Zip II TFs during plant adaptation under changing environmental conditions; how these TFs integrate with phytohormone-mediated responses; the limits of the current knowledge that relate to the mechanism of transcriptional activity; how to overcome these limitations).

Analysis of end of the Introduction from Wymore et al. (2011)

1 Option(s) present in Stage 4: a. Statement of aim or purpose of the present work; b. Statement of main activity of the paper; c. Summary of the findings or outcomes of the paper.
2 Words/phrases identifying the option: a. The major goal of this review is to explore . . .; b. We develop our ideas . . .; we explore . . .; we explore . . .; we explore . . .; c. Thus, . . . can broaden our understanding of . . . and remind us of . . .
3 Section functions as map of how article is arranged: Yes (words signalling this function: . . . in conifers, . . . how the interactions of foundation species (trees and squirrels) and climate can affect a much larger community; how a single mutation in one example and a single haplotype in another example can have cascading effects to redefine their respective ecosystems; how pollution can alter the gene expression of foundation species, which, in turn, may redefine these ecosystems).

The highlighted words sell the novelty and significance of the manuscript to the editor:

On behalf of *co-authors* I am pleased to submit the manuscript entitled <u>"Remnant woodland biodiversity gains under 10 years of revealed-price incentive payments"</u> for publication in the *Journal of Applied Ecology*. This study addresses a major challenge of applied ecology; the evaluation of outcomes from "at scale" conservation incentive program (Payments for Ecosystem Services program). Our study provides the first example of use of a Before-After-Control-Impact monitoring design to evaluate the effectiveness of incentives for restoration of remnant native vegetation. Additionally, the evaluation of conservation changes is in relation to a program where landholders set their own prices for management and were contracted to undertake management prescriptions over a ten-year period. The study provides evidence that management of remnant native vegetation under a revealed-price incentive program can produce gains in vegetation condition.

We believe the manuscript is very well positioned within the scope of the *Journal of Applied Ecology* and work published in your journal over the past few years. This study provides important evidence for contributors to conservation program effectiveness for managers and policy makers, particularly those working in agricultural landscapes.

The manuscript has been prepared according to the instructions to authors and we hope that the information supplied meets your requirements. No part of this manuscript has been published previously or is under consideration for publication elsewhere. All authors agree to the content of the manuscript.

We thank you in advance for the time it takes to consider our manuscript and we look forward to receiving your response and the feedback of reviewers.

Task 17.1 Drafting a sentence template for Stage 4 of an Introduction

Kaiser et al. (2003)

In this study we have identified [NP1], [NP1a]. We show that [NP1a] is [NP2], expressed in [NP3] at [NP4], and is localised to [NP5]. [NP1a] is capable of [NP6] when expressed in [NP7].

Britton-Simmons & Abbott (2008)

In this study we used [NP1] as a study system to better understand the effects of [NP2] and [NP3] on [NP4]. In a [adjectives] experiment we manipulated both [NP2] and [NP3] in order to examine how these factors [adverbs] influence [NP5] in [NP6]. We supplement the experimental results with [NP7], which we use to examine how [NP8] influence [NP9] in [NP10].

Ganci et al. (2012)

Here we present [NP1], named [NP2], which integrates [NP3] with [NP4] to simulate [NP5] . . . Here we describe and demonstrate the operation of [NP2] using a [adjective] analysis of [NP6].

Task 17.5 Generic noun phrases

<u>Legumes</u> form <u>symbiotic associations</u> with <u>N2-fixing soil-borne bacteria of the Rhizobium family</u>. The symbiosis begins when <u>compatible bacteria</u> invade <u>legume root hairs</u>, signalling the division of <u>inner cortical root cells</u> and the formation of <u>a nodule</u>. <u>Invading bacteria</u> migrate to the developing nodule by way of <u>an "infection thread,"</u> comprised of <u>an invaginated cell wall</u>. In the inner cortex, <u>bacteria</u> are released into the cell cytosol, enveloped in <u>a modified plasma membrane</u> (the peribacteroid membrane (PBM)), to form <u>an organelle-like structure</u> called the symbiosome, which consists of bacteroid, PBM[a] and the intervening peribacteroid space (PBS; Whitehead and Day, 1997). The bacteria, subsequently, differentiate into the N$_2$-fixing bacteroid form. The symbiosis allows the access of <u>legumes</u> to <u>atmospheric N2</u>, which is reduced to <u>NH4±</u> by the bacteroid enzyme <u>nitrogenase</u>. In <u>exchange</u> for <u>reduced N</u>, the plant provides <u>carbon</u> to the nodules to support bacterial respiration, a low-oxygen environment in the nodule suitable for <u>bacteroid nitrogenase activity</u>, and all the essential nutritional elements necessary for <u>bacteroid activity</u>. Consequently, <u>nutrient transport</u> across the PBM is <u>an important control mechanism</u> in the promotion and regulation of the symbiosis.

[a] N.B. These two nouns could be considered as countable here, but the authors have used them as uncountable, referring to tissue types. The use of articles is one of the most difficult areas of English grammar, and there is considerable debate about particular cases, even by so-called experts. Use a concordancer with a discipline-specific corpus to check usage in your research field.

Task 17.6 Specific noun phrases

The specific noun phrases are shown with grey background.

<u>Legumes</u> form <u>symbiotic associations</u> with <u>N2-fixing soil-borne bacteria</u> of the Rhizobium family. The symbiosis begins when <u>compatible bacteria</u> invade <u>legume root</u> hairs, signalling the division of <u>inner cortical root cells</u> and the formation of <u>a nodule</u>. <u>Invading bacteria</u> migrate to the developing nodule by way of <u>an "infection thread,"</u> comprised of <u>an invaginated cell wall</u>. In the inner cortex, <u>bacteria</u> are released into the cell cytosol, enveloped in <u>a modified plasma membrane</u> (the peribacteroid membrane (PBM)), to form <u>an organelle-like structure</u> called the symbiosome, which consists of bacteroid, PBM and the intervening peribacteroid space (PBS; Whitehead and Day, 1997). The bacteria, subsequently, differentiate into the N2-fixing bacteroid form. The symbiosis allows the access of <u>legumes</u> to <u>atmospheric N2</u>, which is reduced to <u>NH4±</u> by the bacteroid enzyme <u>nitrogenase</u>. In <u>exchange</u> for <u>reduced N</u>, the plant provides <u>carbon</u> to the nodules to support <u>bacterial respiration</u>, <u>a low-oxygen environment</u> in the nodule suitable for <u>bacteroid nitrogenase activity</u>, and all the essential nutritional elements necessary for <u>bacteroid activity</u>. Consequently, <u>nutrient transport</u> across the PBM is <u>an important control mechanism</u> in the promotion and regulation of the symbiosis.

Task 17.7 Articles and plurals in a science paragraph

Propagule pressure is widely recognized as **an** important factor that influences invasion success. Previous studies suggest that **the** probability of successful invasion increases with **the** number of propagules released, with **the** number of introduction attempts, with introduction rate, and with proximity to existing populations of invaders. Moreover, propagule pressure may influence invasion dynamics after establishment by affecting **the** capacity of non-native species to adapt to their new environment. Despite its acknowledged importance, propagule pressure has rarely been manipulated experimentally and **the** interaction of propagule pressure with other processes that regulate invasion success is not well understood.

N.B. The term *propagule pressure* remains generic throughout the paragraph. It refers to a concept – any or all instances of the concept – and the term *pressure* in this sense is uncountable, so no article is needed. *Introduction rate*, *proximity*, and *invasion success* are likewise generic and uncountable in this passage, so no article is needed.

Task 17.8 Punctuation with "which" and "that"

1 Lime, which raises the pH of the soil to a level more suitable for crops, is injected into the soil using a pneumatic injector.
2 *No additional punctuation required.*
3 Non-cereal phases, which are essential for the improvement of soil fertility, break disease cycles and replace important soil nutrients.
4 Senescence, which is the aging of plant parts, is caused by ethylene that the plant produces.
5 *No additional punctuation required.*
6 Seasonal cracking, which is a notable feature of this soil type, provides pathways at least 6 mm wide and 30 cm deep that assist in water movement into the subsoil.
7 *No additional punctuation required.*
8 Yellow lupin, which may tolerate waterlogging better than the narrow-leafed variety, has the potential to improve yields in this area.
9 *No additional punctuation required.*

Measures of journal impact and quality

There is no easy way to assess the quality of a journal or the contribution of a journal to a research discipline over time. A number of indices have been developed to provide information on the relative speed and volume of citation to journals, and these can give some guidance about their relative popularity and usage. The most commonly used measures of journal impact are the Journal Impact Factor and the CiteScore.

A.1 Journal impact

The *Journal Impact Factor* for a given year is the average number of times articles published in a particular journal in the two previous years have been cited in that year (based on Web of Science data (Clarivate Analytics)). This index provides a measure of the average recent use of articles in that journal. It is calculated using the following formula:

$$\text{Journal Impact Factor}\left(\text{Year}_x\right) = \frac{\text{Cites in Year x to recent articles}\left(\text{Year}_{x-1} + \text{Year}_{x-2}\right)}{\text{Number of recent articles}\left(\text{Year}_{x-1} + \text{Year}_{x-2}\right)}$$

The *CiteScore* (based on Scopus data (Elsevier)) is similar to the Journal Impact Factor but uses a 3-year citation window rather than a 2-year one. The Scopus dataset is also larger and incorporates more social science and humanities journals.

Writing Scientific Research Articles: Strategy and Steps, Third Edition.
Margaret Cargill and Patrick O'Connor.
© 2021 John Wiley & Sons Ltd. Published 2021 by John Wiley & Sons Ltd.

Other metrics of journal quality include:

- *5-Year Journal Impact Factor*, calculated as the average number of times articles published in a given journal in the five previous years have been cited in the current year. The formula is the same as for the Journal Impact Factor, but calculates impact over the last 5 years rather than the last 2 years.
- *Journal Immediacy Index*, calculated as the number of citations to articles in a year with respect to the number of articles published in that year, giving a measure of how rapidly the average article in a given journal is used.
- *Journal Cited Half-Life*, calculated as the number of publication years from the current year that account for 50% of citations received by a given journal, giving a measure of the longevity of use of the average article in that journal.
- *SCImago Journal Rank* (SJR; Scopus), calculated as weighted citations received by a given journal. Citation weighting depends on subject field and the prestige (SJR) of the citing journal.
- *Source Normalized Impact per Paper* (SNIP; Scopus), calculated as the actual citations received relative to citations expected for a journal's subject field. This measure recognizes that some fields of research have inherently lower citation rates than others.

A.2 Using indices of journal quality

Statistics on citation number as a measure of journal quality should be used with an awareness of the purpose for which they are gathered and the limitations of the indices just described. These indices all measure the rate or volume of citation of the *average article* in a journal. They are measures of the journal and not of the individual articles. The number of citations for your article can also be calculated, and it may be higher or lower than the average for the journal. Getting your articles read and cited (or used) is about reaching the right audience. Sometimes, the right audience may not be the readership of the journal with the highest impact factor.

Other things to consider when assessing indices for ranking journals are these:

- Comparing journals from different fields of research may not be meaningful (e.g. mathematics papers cite very few articles, whereas papers in molecular biology journals cite dozens).
- The calculation of some indices is prone to inflating the relative contribution of journals which include contribution types such as comment or opinion (rather than original research).
- Citation-matching procedures are affected by sloppy referencing, the editorial characteristics of journals, some referencing conventions, language problems, author-identification problems, and unfamiliarity with names from some countries.
- Published indices are calculated from a selected database of journals. This database largely excludes journals published in languages other than English and in countries where English is an additional language, and may not include new journals still establishing their reputation.

- Journal ranking based on indices can change over time as research fields evolve, editorial policy changes, and new journals emerge and compete. However, individual articles will largely continue to be cited on their individual merit.
- The ultimate impact of your publications will be determined by many factors other than the impact factors of the journals they are published in. Keeping a healthy perspective on the importance of measures of journal impact will help you develop a career based on contributions of quality research to an increasingly accessible scientific literature.

References

Anthony, L. (2017). AntFileConverter (Version 1.2.1) [Computer Software]. Waseda University, Tokyo. Available from https://www.laurenceanthony.net/software/antfileconverter/.

Anthony, L. (2019) AntConc (Version 3.5.8) [Computer Software]. Waseda University, Tokyo. Available from https://www.laurenceanthony.net/software/antconc/.

Britton-Simmons, K.H. & Abbott, K.C. (2008) Short- and long-term effects of disturbance and propagule pressure on a biological invasion. *Journal of Ecology*, 96, 68–77.

Burgess, S. & Cargill, M. (2013) Using genre analysis and corpus linguistics to teach research article writing. In: *Supporting Research Writing: Roles and Challenges in Multilingual Settings* (ed. V. Matarese), pp. 55–71. Woodhead Publishing, Cambridge.

Burrough-Boenisch, J. (1999) International reading strategies for IMRD articles. *Written Communication*, 16 (3), 296–317.

Cadman, K. & Cargill, M. (2007) Providing quality advice on candidates' writing. In: *Supervising Doctorates Downunder: Keys to Effective Supervision in Australia and New Zealand* (eds C. Denholm & T. Evans), pp. 182–191. ACER, Melbourne.

Cargill, M. (2011) Collaborative interdisciplinary publication skills education: implementation and implications in international science research contexts (Doctor of Education thesis). The University of Adelaide, Adelaide. Available from http://digital.library.adelaide.edu.au/dspace/handle/2440/72719.

Cargill, M. (2019) The value of Writing for Publication workshops. In *Novice Writers and Scholarly Publication: Authors, Mentors, Gatekeepers* (eds P. Habibie & K. Hyland), pp. 195–213. Palgrave Macmillan, Cham.

Cargill, M. & O'Connor, P. (2010) Structuring interdisciplinary collaboration to develop research students' skills for publishing research internationally: lessons from implementation. In: *Interdisciplinary Higher Education: Perspectives and Practicalities. International Perspectives on Higher Education Research Volume 5* (eds M. Davies, M. Devlin, & M. Tight), pp. 279–292. Emerald Group Publishing, Bingley.

Cargill, M. & Smernik, R. (2016) Embedding publication skills in science research training: a writing group program based on applied linguistics frameworks and facilitated by a scientist. *Higher Education Research and Development*, 35 (2), 229–241

Writing Scientific Research Articles: Strategy and Steps, Third Edition.
Margaret Cargill and Patrick O'Connor.
© 2021 John Wiley & Sons Ltd. Published 2021 by John Wiley & Sons Ltd.

226

Cargill, M., Gao, X., Wang, X., & O'Connor, P. (2018) Preparing Chinese graduate students of science facing an international publication requirement for graduation: adapting an intensive workshop approach for early-candidature use. *English for Specific Purposes*, 52, 13–26.

Flowerdew, J. & Li, Y. (2007) Language re-use among Chinese apprentice scientists writing for publication. *Applied Linguistics*, 28, 440–465.

Flowerdew, L. (2016) A genre-inspired and lexico-grammatical approach for helping postgraduate students craft research grant proposals. *English for Specific Purposes*, 42, 1–12.

Ganci, G., Vicari, A., Cappello, A., & Del Negro, C. (2012) An emergent strategy for volcano hazard assessment: from thermal satellite monitoring to lava flow modelling. *Remote Sensing of Environment*, 119, 197–207.

Harris, J.C., Hrmova, M., Lopato, S., & Langridge, P. (2011) Modulation of plant growth by HD-Zip class I and II transcription factors in response to environmental stimuli. *New Phytologist*, 190, 823–837.

Johnson, A.M. (2011) *Charting a Course for a Successful Research Career: A Guide for Early Career Researchers*, 2nd edn. Elsevier BV, Amsterdam.

Kaiser, B.N., Moreau, S., Castelli, J., Thomson, R., Lambert, A., Bogliolo, S., Puppo, A., & Day, D.A. (2003) The soybean NRAMP homologue, GmDMT1, is a symbiotic divalent metal transporter capable of ferrous iron transport. *The Plant Journal*, 35, 295–304.

Kerans, M.E., Murray, A., & Sabaté, S. (2016) Content and phrasing in titles of original research and review articles in 2015: range of practice in four clinical journals. *Publications*, 4 (2), 11.

Kerans, M.E., Marshall, J., Murray, A., & Sabaté, S. (2020) Research article title content and form in high-ranked international clinical medicine journals. *English for Specific Purposes*, 60, 127–139.

Kranner, I., Minibayeva, F.V., Beckett, R.P., & Seal, C.E. (2010) What is stress? Concepts, definitions and applications in seed science. *New Phytologist*, 188 (3), 655–673.

Li, Y., Cargill, M., Gao, X., Wang, X., & O'Connor, P. (2019) A scientist in interdisciplinary team-teaching in an English for Research Publication Purposes classroom: beyond a "cameo role." *Journal of English for Academic Purposes*, 40, 129–140.

Li, F., Zhao, S., & Geballe, G.T. (2000) Water use patterns and agronomic performance for some cropping systems with and without fallow crops in a semi-arid environment of northwest China. *Agriculture, Ecosystems and Environment*, 79, 129–142.

Lindsay, D. (1995) *A Guide to Scientific Writing*, 2nd edn. Longman, Melbourne.

Lindsay, D. (2011) *Scientific Writing = Thinking in Words*. CSIRO Publishing, Collingwood.

McNeill, A.M., Zhu, C.Y., & Fillery, I.R.P. (1997) Use of *in situ* 15 N-labelling to estimate the total below-ground nitrogen of pasture legumes in intact soil-plant systems. *Australian Journal of Agricultural Research*, 48, 295–304.

Methley, A.M., Campbell, S., Chew-Graham, C., McNally, R., & Cheraghi-Sohi, S. (2014) PICO, PICOS and SPIDER: a comparison study of specificity and sensitivity in three search tools for qualitative systematic reviews. *BMC Health Services Research*, 14, 579.

Moher, D., Liberati, A., Tetzlaff, J., & Altman, D.G. for The PRISMA Group. (2009) Preferred reporting items for systematic reviews and meta-analyses: the PRISMA statement. *PLoS Med*, 6 (7): e1000097.

Mullins, G. & Kiley, M. (2002) "It's a PhD, not a Nobel Prize": how experienced examiners assess research theses. *Studies in Higher Education*, 27 (4), 369–386.

Pechenik, J.A. (1993) *A Short Guide to Writing about Biology*, 4th edn. Addison Wesley Longman, New York.

Petticrew, M. & Roberts, H. (2008) *Systematic Reviews in the Social Sciences: A Practical Guide*. John Wiley & Sons, Chicago, IL.

Popay, J., Roberts, H., Sowden, A., Petticrew, M., Arai, L., Rodgers, M., Britten, N., Roen, K., & Duffy, S. (2006) Guidance on the conduct of narrative synthesis in systematic reviews: a product from the ESRC methods programme. Available from https://www.lancaster.ac.uk/media/lancaster-university/content-assets/documents/fhm/dhr/chir/NSsynthesisguidanceVersion1-April2006.pdf.

Sarpeleh, A., Wallwork, H., Catcheside, D.E.A., Tate, M.E., & Able, A.J. (2007) Proteinaceous metabolites from *Pyrenophora teres* contribute to symptom development of barley net blotch. *Phytopathology*, 97, 907–915.

Siddaway, A.P., Wood, A.M., & Hedges, L.V. (2019) How to do a systematic review: a best practice guide for conducting and reporting narrative reviews, meta-analyses, and meta-syntheses. *Annual Review of Psychology*, 70, 747–770.

Tufte, E.R. (2001) *The Visual Display of Quantitative Information*. Graphics Press, Cheshire, CN.

Tufte, E.R. (2020) *Seeing With Fresh Eyes: Meaning, Space, Data, Truth*. Graphics Press, Cheshire, CN.

Weissberg, R. & Buker, S. (1990) *Writing Up Research: Experimental Research Report Writing for Students of English*. Prentice Hall Regents, Englewood Cliffs, NJ.

Wymore, A.S., Keeley, A.T.H., Yturralde, K.M., Schroer, M.L., Propper, C.R., & Whitham, T.G. (2011) Genes to ecosystems: exploring the frontiers of ecology with one of the smallest biological units. *New Phytologist*, 191 (1), 19–36.

Index

Writing Scientific Research Articles: Strategy and Steps, Third Edition.
Margaret Cargill and Patrick O'Connor.
© 2021 John Wiley & Sons Ltd. Published 2021 by John Wiley & Sons Ltd.